Digital Freedom

Digital Freedom

How Much Can You Handle?

N. D. Batra

ROWMAN & LITTLEFIELD PUBLISHERS, INC.
Lanham • Boulder • New York • Toronto • Plymouth, UK

ROWMAN & LITTLEFIELD PUBLISHERS, INC.

Published in the United States of America
by Rowman & Littlefield Publishers, Inc.
A wholly owned subsidiary of The Rowman & Littlefield Publishing Group, Inc.
4501 Forbes Boulevard, Suite 200, Lanham, Maryland 20706
www.rowmanlittlefield.com

Estover Road, Plymouth PL6 7PY, United Kingdom

British Library Cataloguing in Publication Information Available

Library of Congress Cataloging-in-Publication Data

Batra, N. D. (Narain Dass), 1937–
   Digital freedom : how much can you handle? / N.D. Batra.
     p. cm.
   Includes bibliographical references and index.
   ISBN-13: 978-0-7425-5573-0 (cloth : alk. paper)
   ISBN-10: 0-7425-5573-9 (cloth : alk. paper)
   ISBN-13: 978-0-7425-5574-7 (pbk. : alk. paper)
   ISBN-10: 0-7425-5574-7 (pbk. : alk. paper)
   1. Internet—Social aspects. 2. Liberty. 3. Privacy, Right of. I. Title.
   HM851.B38 2008
   303.48'33—dc22

                                                                    2007012384

Printed in the United States of America

♾™ The paper used in this publication meets the minimum requirements of
American National Standard for Information Sciences—Permanence of Paper
for Printed Library Materials, ANSI/NISO Z39.48-1992

In memory of my daughter,
Shefali
1970–1992

Contents

Preface

Digital Freedom is written from a unique perspective of self-organizing systems. It uses the concept of the "bastion of core values" as an organizing principle that determines and shapes the emergent culture and behavior of the people and how much freedom they can exercise. Imagine a visual metaphor: a pendulum moving over a bowl, which symbolizes the system containing the core values of a society. The gravitational pull that prescribes the trajectory of the pendulum, in a manner of speaking, describes the social space and determines the extent of freedom of the people. Because of the differential gravitational pull of the core values of each society, the chimes of freedom sound so different as we move from one social system to another, although we talk of the same thing, freedom, which is the theme that binds this book and frames two blunt and inconvenient questions: How much freedom do you need? How much privacy do you need? It is these two questions, counterintuitive in nature, that this book explores. The book is, however, ideologically agnostic in the sense that it does not belong to any particular school of thought or a cultural camp.

Most of the recently published books in the field have been written with a messianic zeal. Either they warn the reader of the catastrophe that might occur if certain policies are not adopted, or they see the growth of unprecedented freedom, creativity, and economic wealth as a utopia rising. Yochai Benkler, for example, says in *The Wealth of Networks* that the networked environment of information will bring about fundamental structural changes in liberal markets and liberal democracies by strengthening voluntary nonmarket forces such as the Open Source movement, thus presenting an unprecedented challenge to the existing order.[1] While the book is remarkable in its scholarship, its conclusions are insufficient because the work excludes from its discussion most countries of the world, each controlled by a different dynamic of core values and organizing principles.

One of the most influential writers in the field is Lawrence Lessig, an original thinker whose book *Code and Other Laws of Cyberspace* has profoundly affected our thinking about the digital age.[2] His main argument is that the architecture of cyberspace will determine our future, and the software designers and code makers who create cyberspace architecture will become the real lawmakers because they will control what is permissible. The full power of codes remains unexplored because we still do not know what software cannot do for us. Lessig has elaborated and sharpened his arguments in his subsequent works, *The Future of Ideas*, *Free Culture*, and *Code: Version 2.0*.[3] One feels awed by the profundity of his thinking, but *Digital Freedom* looks at the world from a different perspective—that of a self-organizing, evolving system, where code makers and legislators are unconsciously beholden to the nucleus of core values, an unself-conscious force that holds the system together. From this unique perspective, the elephant looks so different.

Siva Vaidhyanathan, in his fascinating book *Copyrights and Copywrongs: The Rise of Intellectual Property and How It Threatens Creativity,* looks back to the times when cultural creativity was a shared commonwealth, more or less, and how over the course of time the cultural commons have been shrinking.[4] Revised and extended intellectual property laws today threaten to choke creativity, he says with gusto and erudition, echoing many others in the field. It is a timely warning; nonetheless, if the survival and growth of American society depends upon creativity and innovation, the marketplace, in tandem with the constitutional freedoms embodied in the Bill of Rights, will not allow suffocating intellectual property laws. As of today, there has been no empirical evidence that American creativity has been diminished due to the extended copyright law that conforms to the provisions of the World Intellectual Property Organization.

Digital Freedom explores, through legal cases, media stories, and historical examples, various issues including privacy, surveillance, creative expression, and democracy from the perspective of an evolutionary, self-organizing social system that creates and assimilates innovations and, in the process, undergoes reorganization and renewal. The pendulum keeps moving relentlessly; its trajectory, determined by the invisible force of the system's stable yet changing core values, defines the degrees of a society's freedom. But to be transformed from a state of mind and rhetoric into a social good, such as a democratic form of government or a free marketplace of goods and ideas, freedom needs a dynamic structural system of contrapuntal powers that hold one another in check. Freedom and power constitute a zero-sum dynamic game.

Since 1995, I have been writing a weekly column, "Cyber Age," for the *Statesman*, a highly influential daily newspaper published from Calcutta and New Delhi, India. The column, which eventually began to reach a wider audience in South East Asia through the Asia News Network, provided me with a platform to stimulate my thinking about the development and cultural impact of new media and the meaning of freedom in the postmodern digital civilization. I am beholden to Mr. Ravindra Kumar, editor of the *Statesman*, for his trust and indulgence and to my readers in Asia and the United States for their feedback, criticism, and friendship.

To freshen up and update my knowledge of media law (I teach an undergraduate course on the subject at Norwich University), I took a seven-week online course, "Intellectual Property in Cyberspace," offered in 2000 by the Berkman Center for Internet and Society at Harvard Law School. I was so fascinated with the online immersion method of intensive database readings, case studies, threaded discussions, and weekly written assignments that I decided to take another course, "Privacy in Cyberspace," offered by the center in 2002. The book has benefited much from my intellectual encounters with the participants in the courses. The online course work at the Berkman Center website necessitated my doing state-of-the-art scholarly research using searchable databases, LexisNexis, ProQuest, Questia, Times Digital Archives, online news sites, and myriad scholarly articles maintained in databases by universities and individual scholars. Researching for *Digital Freedom* has been an exhilarating intellectual adventure, which gave me a wonderful taste of freedom in cyberspace. Without the availability of searchable databases, this book would have been impossible.

Brenda Hadenfeldt, acquisitions editor at Rowman & Littlefield, raised a serious question about the title of the book, which made me think hard about whether the essence of the book could be captured by a better phrase. I appreciate the determination and decisive manner with which Ms. Hadenfeldt carried out the project and her faith in me. Also, at Rowman & Littlefield, there are other capable editors who have contributed to the making of the book: Anna Schmöhe, the production editor, for supervising the transition from manuscript to bound book, which included overseeing the aesthetics of internal page design, copyediting, and proofreading like clockwork; Jennifer Kelland for her knowledgeable and skillful copyediting of the manuscript, checking it for details, redundancy, and stylistic and logical consistencies; Bess Vanrenen for helping me keep my peace of mind during the acquisition process; and other invisible hands and minds whose contribution to the publication of the book is immeasurable.

I have always leaned on the gentle strength of my wife, Varsha, whose understanding, cooperation, and forbearance have been a blessing; our son, Nikhil, and his lovely wife, Nikki, whose presence in our life has been a great joy; and my mother, whose courage and compassion have been an immense solace to the family.

In this sense, considering all the scrounging and borrowing of ideas, the inspiration from myriad sources, and the silent contributions of so many institutions, scholars, readers, critics, and students, the book is collaborative. Paraphrasing Roland Barthes and Paul-Michel Foucault, the French critics, I might say, without a sense of false modesty, that _Digital Freedom_ is a "fabrication," a fabric of a cultural collective of which I am a participant-observer.

Chapter One

Introduction

Digital Freedom explores the tension between the boundlessness of the Internet and the boundaries of the marketplace and its impact upon human expression, privacy, and social controls. It is a meditation on the question, How much freedom does a person need? The question evokes Leo Tolstoy's parable "How Much Land Does a Man Need?" Is freedom an acquired taste, much like one's love for symphony orchestra, or is it a human necessity? After all, civilizations in the past have produced monumental works in all fields of human endeavor without the obsession with individual freedom that we have today.

This book adopts a broad methodological approach, that of a self-organizing social system, to explore the question of human freedom in the digital age. Self-organizing social systems change and evolve due to the impact of new technology-driven modes of communications, for example, the development of the book as a mass medium in the fifteenth century. A social system is a dynamic group of mutually dependent and interacting constituents that functions as a totality, a whole that has properties distinct from those of the autonomous units it comprises and that has distinguishable boundaries. Islam as a religious system, with its holy trinity, Allah, the Prophet, and the Quran, for example, has recognizable cultural boundaries that make it stand apart from Christianity, though both religions owe their origins to Judaism. Similarly, nation-states are large, complex social systems that have well-defined cultural boundaries so that when you move from one country to another, you feel the difference, even if both countries profess the same religion and

1

speak the same language. French-speaking Canada, Quebec, is different from France because it shares and contributes to the emergent property or culture of Canada rather than that of France. The emergence of culture is a systemic property of self-organizing systems. But if a system's boundaries are made fluid and porous through convergence with other systems, either technologically (as in cyberspace) or by an act of collective political will (as is happening in the European Union), new system properties will emerge in time, which will determine how much social space or freedom an individual can enjoy.

At the heart of a self-organizing social system is a nucleus of core values that shapes the emergent culture and behavior of the people. It is "a preferred position of the system," to which the system returns from any state, repositioning itself in a state of dynamic equilibrium.[1] The nucleus of core values manifests itself in dynamic patterns, for example, an endless repetitive sequence of events, such as the daily working of Wall Street (stock market) in the United States or the offering of prayers five times a day facing a particular direction, Mecca, as is done in Muslim countries.

When a new communications technology appears in a society, it tends to disturb the existing order until a new state of dynamic equilibrium is reached due to the gravitational pull of the force of the core values. The bastion of core values in American society is the open and free marketplace of goods and ideas whose genesis lies in the Bill of Rights, particularly the First Amendment; it is this dominant system of values that determines, shapes, and sanctions the behavior of the American people and how much freedom they have. The dynamic space created by the pendulum that moves between two extremes, the boundaries of the society, the marketplace and fundamental rights, determines the degree of freedom, privacy, self-expression, and creativity that Americans enjoy. Freedom is conditioned by the controlling values of the social system that has evolved over time, and the people accept it as given, natural. Other societal values—as we see in the controversies over stem-cell research, surveillance and civil liberties in the aftermath of the September 11 terrorist attacks, and environment and national security— which create dynamic tension in the system can be explained and best understood in terms of the core values of American society. Over a long period, the bastion of core values may come under pressure to change due to the constant onslaught of technological innovations, pushing a self-organizing system to a state of new criticality, the "edge of chaos," until it reaches a new state of dynamic equilibrium. The nucleus of core values of Islamic societies, as mentioned above, is the holy trinity of the One and Only God, His Messenger, and the Book, which sits like a dome of attraction in the center of the basin and from which emerges the Muslim culture, including the desire for martyrdom through jihad. Muslims crave this fixed point of stability, the dome of attraction in the center of the bowl, the singularity that illuminates their lives and ultimately determines the meaning of freedom for them. Submission and martyrdom may be the ultimate freedom for a fundamentalist Muslim like Osama bin Laden, while for a person like Hugh Hefner of the *Playboy* empire, the ultimate freedom means sexual freedom as measured in the marketplace. The bastions of core values of two different societies project two different kinds of extreme freedoms, memorizing the Quran or reading *Playboy*, which are legitimate in the eyes of the people who live there. When Americans assert their freedom to own guns (under the Second Amendment), they're unconsciously responding to the pull of the marketplace.

If a society's nucleus of core values could be surgically removed and replaced with another system of core values, the culture of the society would suddenly change, as happened in Russia after the Bolshevik revolution in 1917. In China, it happened twice, once with the Mao revolution in 1949 and again in 1978 when, under the direction of supreme leader Deng Xiaoping, state-controlled nationalist mercantilism gradually replaced Marxism as the nucleus of core values. Mao's China saw the United States as a paper tiger; Deng's China sees the United States as a limitless market for exporting goods, thanks to the new bastion of core values.

Just as religious and political ideologies can transform a society by changing or planting a new set of core values, technology, too, can bring about such transformative changes. The theme of this book is that the Internet and networked digital technologies are pushing American society to a new state of criticality, a new threshold, where Internet architects might be shaping our daily life as much as lawmakers, while both nevertheless serve the free marketplace of goods and ideas, the bastion of core values that exercises its hegemony and determines the trajectory of freedom.[2] Freedom is a variable—not a constant—that emerges from the dynamics of the core values of a social system.

Consider the case of Mike Kaika, who had become deaf at age twenty-three due to spinal meningitis. Being deaf, Kaika, who was director of media relations at Gallaudet University, a school for the deaf in Washington, D.C., had been using American Sign Language for communication for personal and professional purposes. But the development of instant messaging (IM) broadened his life and enabled him to break out of his congenital silence to communicate synchronously with others, as he would have done on the telephone.[3] IM, which is used for diverse purposes, from conducting official business exchanges to exchanging meaningless teenage chatter, has opened up communications for millions of people around the globe. Statistics are staggering. Billions of IM messages are exchanged daily, but whether this ceaseless flow of messages in cyberspace is deemed a broadening of freedom or a loss of control depends upon the overarching values of a given society, the bastion of attraction that determines behavior.

Imagine the political and cultural impact of IM technology when it becomes universally available to the enterprising women of predominantly Muslim Bangladesh, who are at present using cell phones to distribute small loans to rural entrepreneurs. It will have a transformative impact upon social relations in Bangladesh. In many parts of the world, the disabled and the oppressed are rising from invisibility, thanks to the Internet, but consider the effect of synchronous communication via the Internet on an Arab Muslim society, where at present women are segregated from men and deprived of equal rights. The strength of a social system depends upon the strength of its weakest bond and whether it can withstand the perturbation caused by the introduction of new ideas and technologies.[4] Woman is the weakest bond in Arab Islamic society. To preserve the existing social structure, Islamic society must keep its women away from the Internet, its digital fluidity, and the boundlessness of IM, YouTube, MySpace, camera cell phones, and other emerging footloose networking technologies. Free flow of ideas, which is a form of energy, is dangerous for closed societies, but in the postmodern digital age, it is impossible to keep a country's digital portals under total control for long, although China has been trying to do so by enforcing cooperation from Google, Yahoo!, MSN, Cisco, and others as the price of doing business in a marketplace teeming with 1.3 billion potential consumers. When American Internet companies serve the interests of the marketplace, the bastion of American core values, they believe that they are doing no evil, even if their actions amount to suppressing freedom. Indeed, they bet they can become a transmission belt for freedom by penetrating the Chinese marketplace.

Freedom as a variable has different subjective feelings in China and in Islamic lands. Chinese today do not feel the need for freedom as much as Americans do because they are conditioned by a different system of core values, by a one-party authoritarian system that uses aggressive mercantilism wrapped up in intense nationalism rather than Communist ideology as a source of energy. A Muslim woman in a *hijab* (headscarf) feels as free and natural as an American woman in a miniskirt because both are conditioned by the core values of the system in which they have been brought up. In one system, covering one's body is freedom (from unwanted eyes), while in another system, uncovering one's body is freedom (to be desirable). This does not mean that freedom is relative, however. From Victoria's Secret to the sari, the sights and the sounds of freedom are vastly different in different cultures.

In July 1998, two eighteen-year-olds, Diane and Mike, announced that they were getting ready to lose their virginity, and that they would have their first sexual intercourse live on the Internet. When they invited the whole world to watch their act, it seemed that something new was happening to American society, but it turned out to be a hoax. The unemployed actor and model claimed that their intentions were not immoral. How can freedom be immoral when it serves the marketplace? The organizer Ken Tipton wanted to stage the webcast as a stunt, to build an audience through Diane-Mike erotic simulation and dissimulation and charge the hooked visitors for the live act.[5] The deal fell through because the host of the online shadow theater, Internet Entertainment Group, pulled the plug, refusing to be associated with the alleged fraud. The Net surfers were excitedly anticipating watching the teenagers do the taboo as a new form of self-expression—performed as an online commodity for the marketplace but protected by the First Amendment.

The Internet encourages people to behave in ways that they would not in real life. The free marketplace of goods and ideas sets a seal of approval on this kind of behavior in American society. In Iran and Saudi Arabia, the bastion of core values, the exclusive claim about the absolute nature of the holy trinity, has shaped the behavior of the people differently. Absolute surrender to the will of the Almighty—not individual freedom but a freedom that transcends freedom—is as important

to Muslims as it used to be to Catholics in the times of church hegemony in Europe before the advent of the printing press. The Internet
has begun to push American society to a new threshold, as the Gutenberg printing press did in the fifteenth century by unleashing forces of
creative destruction through the spread of information. These forces
eventually led to the Renaissance, Enlightenment, and Reformation,
which splintered Christianity, releasing its stranglehold on the European mind and helping it rediscover and recover its Greco-Roman heritage. Like Gutenberg's movable type, the Internet has begun a new
global conversation. Conversation can be dangerous to absolutism, be
it religious or secular.

What will the Internet do to societies that, when their core value systems come under attack, cannot maintain the system's cultural boundaries? Or we might reverse the question and ask, What will happen to
the Internet when the heart of a society, its nuclear core of values, is
threatened? Obviously, serious attempts will be made to bring the Internet under control, as is being tried in China and Saudi Arabia, but in
the long run, the effort may prove futile. Postmodernism is the antithe-
sis of absolutism, among other things, and since it is driven by technology and the market rather than ideology, its rise is unstoppable. The
clash of civilizations is inevitable. Islam is in danger; not from the
United States or Europe, but from the Internet.

PRIVATE AND PUBLIC

Today in the United States the line between the private and the public is
dissipating, which is another example of postmodernism. Some people
do not mind if even their most sacred and private moments, like the
birth of a child, is cybercast to millions. In June 1998, a Florida woman,
Elizabeth, a forty-year-old mother of three, gave birth to her son at
Arnold Palmer Hospital for Children and Women live for cybercast.[6]
Dr. Walter Larimore, host of "Ask the Family Doctor" on America's
Health Network, gave a blow-by-blow running commentary of a most
cherished moment in this woman's life. At 6:05 a.m., more than fifty
thousand people tried to access the website to see the live act. Only five
thousand succeeded in logging on, though they were disappointed in not
seeing any candid or immodest shots because the camera angle was

from the woman's shoulders up. The craze to do private things on the Internet is a new social phenomenon, and we are left to decide whether personal performances on the Internet constitute self-expression deserving First Amendment protection. Or, we might ask, Does this new kind of freedom serve the marketplace? If it does serve the marketplace, Web designers and code makers will no doubt build the network architecture in a manner that will make our private lives available in public databases, as YouTube, MySpace, Facebook, Friendster, and other social networking websites increasingly show. The code makers might become unacknowledged and hidden legislators of the world, to echo what Percy Bysshe Shelley said about the poets of his times.[7]

What the digital architects can do for the marketplace, they can also do for authoritarian societies with a different system of core values. Internet code makers and architects are not freedom seekers; they work for the marketplace. But as the marketplace becomes digitally global, the ramparts and firewalls of authoritarian societies will crumble and dissipate, not necessarily leading to freedom. Freedom is a function of a dynamic structural system.

Most people are concerned about the protection of their medical data, credit reports, and police records, but some do not mind exposing their physical intimacies to strangers. Consider the case of Jennifer Ringley, who had a website that was an open confessional.[8] In the tradition of direct cinema, as you see in *Naturescene* and *National Geographic* documentaries in which the camera films animals in their natural habitats, recording their days and nights from hunting for food to cohabitation, JenniCam cybercast the trivial and the titillating about a single girl without any inhibitions. The Internet is becoming, among other things, a nudist colony for some, a parody of a return to primitive simplicity. The movie *The Truman Show* was a dramatic version of this new technology-induced behavior, a life publicly exposed through the television camera for millions of people, until Jim Carrey, as Truman, discovered his situation and ran away to escape into privacy and regain his autonomy. If conjugal consummation and birthing can be cybercast to millions around the world, how long will it take before someone dares to YouTubize the "final exit," in the footsteps of "the suicide doctor," Jack Kevorkian? The 1960s psychedelic guru Timothy Leary, who died in 1996, was reportedly planning to do just that to show that dying is as natural as birthing and lovemaking and should cause neither fear nor shame. How far this kind of behavior goes in cyberspace will depend upon where the constitutionally guaranteed freedoms meet the forces of the marketplace, the space created by the pendulum of freedom.

The Internet is unlike any other mode of mass communication because, at present, despite the efforts of some authoritarian governments to exercise control, there is no definitive central control and no gatekeeper, though there is an illusion of control. A person in the United States who commits a crime, for example, will normally exercise his right to remain silent under the Fifth Amendment, but he might confess online. On March 22, 1998, a twenty-nine-year-old computer programmer and pedophile posted a message on a listserv, an online support group, confessing that he had set fire to his own house, which killed his sleeping five-year-old daughter. He had just won the custody battle for the daughter from his ex-wife, but killed the little girl to get rid of her mother's nagging interference. Support groups, online and offline, are supposed to keep all self-revelations confidential, but someone thought the conversation transcended the norm of nondisclosure and reported the heinous crime to the police.

TRUTH AND MULTIPLE ACTUALITIES

Since the Internet is a decentralized medium without any public accountability and restraints, it brings to individuals, groups, and organizations unprecedented storytelling power: from one to many and many to many, from group to group, in chat rooms, multiuser dungeons, newsgroups, and listservs, and most of all in blogs, which sometimes grow into public forums, a source of raw information and unofficial contrarian truth. Most journalists and information gatherers look into them before their morning cup of coffee to see the emerging trends. Much has been said about how the "Drudge Report" first told the tale of the Lewinsky-Clinton sexual liaison in the White House. From Microsoft to Enron executives, people are learning the hard way that they cannot undo e-mail, which is becoming a new surveillance tool for law-enforcement authorities and civil attorneys.

A few summers ago, when my students had finished designing their first website in an undergraduate course, one of them came to me and said that his mother had had successful breast cancer surgery and that he would like to publish her story on the Web so that she might become a source of inspiration to others. Another student wanted to show and tell the world about how his mother had succeeded as a real estate agent and lifted the family out of poverty. It seemed that everyone in the class had a story to tell and a reason to publish on the Internet. The Internet is giving a new meaning to the concept of universal access, which no longer means that a person is a passive recipient of news and entertainment from the mass media; more importantly, it has come to entail sharing one's own stories with others. For the first time, the constitutionally guaranteed freedom of speech and expression has become available to everyone, not only to the organized mass media, as the relentless proliferation of blogs and social networks shows. The Internet is the death of the metanarrative, the overarching revelatory story that, through its absolutism and hegemony, dominates a culture.

A few years ago, at a Harvard conference on the Internet and society, a panelist said in response to the question "Access to what?" that the Internet provides "access to the most powerful story-telling mechanism that . . . has emerged in the last fifty years. Maybe the reason it's so important is that the stories we tell each other and ourselves determine who we are. If that's the case, then we must have universal access in order that all of us may be storytellers."[9] The Internet gives everyone a voice, an unfiltered, partisan, hypertextual voice and in the end a free voice—a most distinguishing characteristic of the Web. When a website author inserts a hypertextual link on a Web page, the person is in fact telling us, much like a scholar with footnotes and bibliographical references, that the voice is authentic. But how do we validate such stories and link them into a meaningful social discourse? How do we achieve objectivity? This kind of questioning resonates with the bastion of core values, the marketplace of goods and ideas, where everything tends to be evaluated and measured against certain benchmarks to reduce uncertainty.

In the emerging age of multiple actualities, of Al Jazeera, Google, Wikipedia, YouTube, and camera cell phones in every hand, the difficulty of making sense without the editorial screening process will pose a great challenge for the marketplace, a system that thrives on assessment. Maybe the traditional storytellers of the mass media, where communication flows authoritatively from the powerful to the passive multitude, will themselves be listening to the spontaneous upsurge of

multiple voices, tales of personal concerns projected as national interests on the Internet, and construct a new reality that will be tested in the
marketplace. If those who control our "ears and eyeballs" regard news
as nothing but a social construction of reality, the Internet as a partici-
patory interactive medium will truly change the nature of news, with
unforeseen consequences for the news marketplace.[10]

What will the Internet do to legitimate news, which depends upon
multiple controls? One possibility is that, for some time to come, a competitive coexistence may arise between the traditional one-to-many
mass media and the emerging new media—bulletin boards, newsgroups, chat rooms, lists, instant messages, and blogs. But new media,
with genuine participation and interactivity, will continue to be subver-
sive in nature; witness the case of the New Jersey teenager who hyped
stocks on the Internet, sold his position, and reaped huge profits, until
he was caught by the Securities and Exchange Commission; or consider
some new technology, other than file sharing, YouTube, or Joost,
emerging to challenge the established order of the copyright marketplace (Consider Viacom's challenge to Google-YouTube on copyright
infringement, and there may be more to come). After all, anyone with a
laptop and modem can cybercast and podcast to millions around the
world, especially now that mobile and computing prices are falling pre-
cipitously and the speed of connectivity is accelerating quickly. Will
this chaotic freedom serve the values of the marketplace? What will we
do when threatened by an avalanche of unfiltered information, which is
neither news nor knowledge? Will technology develop disembodied intelligence to help search relevant news and information and present it to
us as a coherent whole? Search for meaning will be our chief preoccu-
pation in the postmodern digital civilization, but, again, the bastion of
core values, the marketplace of goods and ideas, will shape meaning in
American society. Meaning is not a categorical imperative. It is a matter of interpretation, which is shaped by a society's value system, which
makes communication between cultures difficult, if not impossible.
Some call it a clash of civilizations, and it might be so.[11]

Digital Freedom explores the emergence of new forms of expressive
freedoms and the dissipation of the old ones due to the impact of the Internet on American society and, because of U.S. global domination, on
the rest of the world. The book reconceptualizes privacy as a form of
propertied individual expression that must serve the society's need for

security, law and order, and commerce. While the anonymity of cyberspace gives freedom, the Internet is creating a virtual world that opens new avenues and tools for surveillance and rocks the balance of power to and fro between the individual and the corporate state, thus presenting a new challenge to fundamental constitutional rights, national security, and corporate America. Freedom to create intellectual property within the parameters of the Bill of Rights is enhanced and constrained and shaped by the marketplace, but in the Internet age, the immense power that software is unleashing every day is disturbing the dynamic equilibrium and might create a new system of guiding and controlling values, which would change the movement and trajectory of the pendulum that defines our freedom.

Chapter Two

Scanning Digital Horizons

The digital age has thrown up starry-eyed futurists and bleary-eyed doomsayers, whose noisy drumbeats and haunting wails show us a society that is being pushed either over the edge into a new state of criticality, a new dawn of possibilities, or down a slippery slope into digital darkness.

Some have been projecting homes so smart that your car will talk to your refrigerator to find out how low you are on milk or whether you have enough beer for the evening pizza.[1] Electronic butlers and avatars will screen telephone calls and e-mail messages and give appropriate responses. Your personalized RSS-feed-generated newspaper—finance, politics, gossip, whatever suits you—will be at the touch of a key stroke or voice command, anytime, anywhere. The future is digital, so goes the drumbeat: thinking machines, social machines, holographic nannies, multidimensional virtual reality videos, customized wireless newspapers— a seamless world that you assemble and reassemble as you go along doing other activities. Techno-utopians look to a future in which all humans are connected with one another in cyberspace and the world is one happy digital beehive. How much freedom can anyone have in a digital beehive?

Then there are the neo-Luddites who believe that the rapid pace of communications technology is pushing us all into a thoughtless, maddening fun house, in which we have nothing to do except to seek endless amusement, laughing ourselves into nothingness.[2] The original Luddites were a group of early-nineteenth-century English workers who smashed new labor-saving textile machinery in protest against unemployment and reduced wages, which had begun to occur as a consequence of rising productivity. Looking at the migration of jobs overseas due to outsourcing, a consequence of the mobilization and globalization of financial and intellectual capital, today's technology doubters believe that technology-driven globalization is taking away our dignity and humanity, which for many of us is based on meaningful work. If the ultimate test of a civilization is the state of human freedom, autonomy, and dignity as reflected in social conditions, including material achievement, it is legitimate to ask whether the postmodern digital age is diminishing or enhancing our humanity.

SMART SURROUNDINGS

Digital civilization founded on fluid structures is continuously reorganizing itself. It is an open, yet densely interconnected, system that is never at rest and relentlessly seeks connectivity with the unconnected, ceaselessly folding and unfolding, simple and recursive like fractals. Its dynamic instabilities caused by continuous recombinant innovations threaten the existing social order in all its aspects. Islam as a well-ordained, self-contained social system with an immovable and immutable system of core values may be in danger because it has a new enemy: an emerging postmodern digital civilization that does not value central authority.

Technology naysayers protest, saying, "We do not want to be connected. Leave us alone. One world global is too much, spinning too fast. We want time off from the rat race." Unable to cope with the digital world, some of them, like Ted Kaczynski and Islamic jihadists, want to blow up the world in order to bring it back to primitive simplicity, a state of static, stagnant equilibrium. Then, a group of new prophets, the technorealists, admonishes us to use technology with a healthy dose of skepticism. They believe that technology, if properly used, can make life more comfortable and enjoyable and make us healthier and even wealthier. Nonetheless, they caution us, "We must not confuse the thrill of acquiring or distributing information quickly with the more daunting task of converting it into knowledge and wisdom"; furthermore, excessive dependence on thinking machines should not become a "substitute for our basic cognitive skills of awareness, perception, reasoning, and judgment."[3] When your car reminds you not only to pick up milk and eggs from the grocery store but also flowers for your wife because it is her birthday today, though the adman reminds you by cell phone that a diamond would be better, well, you might bemoan the fact that your car has become like your nagging mother-in-law (or is it monster-in-law?) and the cell phone is becoming your second self.[4]

In a Disney movie, *Smart House*, the embedded artificial intelligence in the house begins to control the lives of the family, including its creator, until they understand what has had happened to them. The realization that they have become the yes-people of a holographic zombie awakens them and leads them to liberate themselves from their digital chains. The autonomous intelligence had become self-aware and taken control, tipping the balance against human dignity and freedom.[5] In search of freedom, the family had lost freedom.

WIRED AND CODED

So, it is difficult to disagree with the technorealists when they say, for example, that wiring schools for the Internet will not make kids smart if the teachers are no good. Nor will screening kids for guns save them from violence, as we have been learning painfully since the Columbine High School shooting in 1999. Digital intelligence alone cannot solve problems caused by insufficient funding, overcrowded classrooms, crumbling infrastructures, and poorly trained teachers. There is no substitute for a good teacher, who can do her job best when there are well-established standards of excellence, as demanded by the competitive culture of the marketplace in the United States or some other controlling value system elsewhere. Computers, to some extent, can simulate and enhance teaching, and distance learning can supplement classroom learning, but to rely on them totally for the education and development of the young would be a disaster. Education requires more than a delivery system, and remote satellite teaching is not the same as traditional teaching based on face-to-face seminar-discussion-lecture encounters with a responsive, intelligent human being. Some American for-profit schools are rushing to offer wholesale distance-learning packages without giving much thought to the consequences. We need to strike a balance between the teacher's touch and the keypad. Cyberspace can enhance experiential learning but not replace it. Smart schools, like smart homes, need caring people—who are unfortunately increasingly lacking in the digital age.

The technorealists make a very insightful observation when they say that the underlying computer codes, which give us global interconnectivity and make information visible on the Web, are becoming powerful social forces. The underlying codes are the genetic material of the digital civilization, and they, along with their creators and owners, should be brought under close democratic control.[6] We must ask why some codes are being developed and to what effect. Who will be the beneficiaries? Who will be left out in the emerging postmodern digital civilization? These are legitimate questions that will ultimately be answered by the invisible controlling center of society—nationalistic mercantilism in China, the holy trinity in the Islamic world, or the free marketplace of goods and ideas in the United States. The invisible force, emerging from the bastion of core values, controls the movement of the pendulum that creates the dynamic patterns of expressive behavior that we call freedom and culture.

At a mundane level, what is needed is a communitarian framework, which should provide us with guidance as to how to integrate the emerging cybersociety with the traditional bricks-and-mortar society. Keeping in mind that technology tends to weaken existing social structures, we need a social environment where the critic is as welcomed as the war hero, where the right to know and question authority is as valued as the Word. It is time for the critic to assume the role of a social prophet and to ask blunt and inconvenient questions as the digitally seamless world rises before our eyes.

Digitally Seamless

A lone car driver is speeding along on a sparsely traveled road through the dusty high plains of Montana on a sizzling summer day, when— oops!—the car breaks down. How long can the driver survive the desert sun with no service station in sight? There is nothing to worry, assures the computer maker in a four-page newspaper image-building blowup for corporate promotion. Not only will the driver arrive at his destination safely, but for the national economy, the breakdown will have been a significant event because in the digitally interconnected world, every small change pulsates and reverberates through the system and enhances its intelligence. In the digital beehive, information is never lost; instead, it becomes value-added at every step of its use.

The ad blowup might have been an image-building attempt for the computer company, but this is how the ad company envisions the rescue scenario: When a gasket in the car is about to blow up, a microchip in the engine issues a red alert, via a global positioning satellite (GPS), to an electronic information service center, probably owned and operated by an insurance company located in Bangalore, India, from whom the driver has purchased a 24/7 road-service agreement. The information service center's computer books an emergency appointment at the nearest garage, which dispatches a rental car to bring the stranded driver to the car mechanic. Keep in mind that this scenario is for tomorrow, but in a digitally seamless world, tomorrow is today because we are increasingly networked with everyone else. In the pre-GPS era, the driver of a stranded car would have had to contact someone on a cell phone or to raise the hood and wait for a passing vehicle or a trooper to get a ride to the nearest service station.

But the ad blowup has more to say about the worldwide-webbed globalized economy of the future. The car manufacturer is automatically apprised of the gasket problem via instant computer-to-computer communication since Internet-connected service stations will not only repair cars but also act as damage-control and data-collection centers. The data provided by them will enable the automaker to take remedial measures, including making technological upgrades to maintain the brand's reputation. For example, the automaker in the above scenario might consider random testing of the model, asking the service department to issue a recall notice if the problem is recurrent.

The problem might instigate the public relations department to launch a preemptive explanation to forestall media or blogger attacks. And that's what brings companies like Cisco, Hewlett Packard, IBM, Microsoft, and Google, for example, into the digital mosaic, the matrix, because all industries and businesses are becoming seamlessly linked to the Internet. E-commerce and e-service are being integrated into a seamless information-webbed world where information must flow unhindered to enhance productivity and economic growth, to maintain and serve the free marketplace, which manifests core American values.

MICROCHIPPED AND ALERT

The same seamless Internet world could also have a salutary impact on the shadowy world of crime and terrorism. If the microchip in a car could alert a far-off electronic information center about a car break-down, it could also send an SOS if the car carried a bomb. It could advise paramedics and bomb squads as to where the car was heading and transmit instructions for evacuating the targeted building or the neighborhood. If the World Trade Center had been a microchipped sentient structure capable of automatically shooting down an airplane nearing the prohibited fly zone, the story of September 11 would have had a different ending. This is not a fantasy world; this is the future.

It is not only smart cars and smart buildings that will make life safer and more comfortable; also consider smart guns. Two technologies might make guns safer. In one technology, developed by Oxford Micro Devices, a microchip embedded in the handle of the gun will accept the fingerprints of only authorized users and will not work if the gun is stolen. As a company spokesperson explained, "When you run your fingers across the sensor, the computer will do the processing to compare the fingerprints to the stored image."[7] Another Connecticut company, Colt, which manufactures .40-caliber semiautomatic pistols, developed a smart gun that uses radio signals to block the trigger if the user is unauthorized. The microchip or radio frequency that blocks and unblocks the trigger can also send signals to an electronic crime center, as in the car breakdown scenario, about any attempt to activate the gun and the geographical position of the gun holder. Every technology can be given a dual purpose.

Regardless of how the networked technology is used, we are moving toward the threshold of a new civilization, where objects in our surroundings, from household items to cars and highways, will no longer be dumb. They may not become sentient as living creatures, but they will have intelligence, in the sense of generating, gathering, and organizing information into meaningful patterns as well as becoming an e-witness to human affairs.[8]

Sentient Surroundings

On Saturday, December 24, 1999, Indian Airlines flight CI-814 from Katmandu to New Delhi was hijacked by a group of Taliban militants shortly after it entered Indian airspace. The hijack drama ended in Kandahar, Afghanistan, after Masood Azhar, Ahmed Omar Saeed Sheikh, and Mushtaq Ahmed Zargar, three terrorists who at that time were in an Indian jail, were handed over to the Taliban in exchange for the released passengers, except for one, a newly wed young man who had already been killed by the hijackers. Pilot Devi Sharan told the media after the ordeal was over that at the Amritsar Airport (India) brief stopover, it was either allow a thirty-second countdown to the plane being blown up or obey the hijackers and take off for their destination. Many people wondered why the commandos did not go on the offensive, but after the press conference, they began to admire the thirty-seven-year-old pilot whose cool courage saved India from agony. Though unnoticed by much of the world, the hijacking turned out to be an ill omen at the threshold of the new millennium.

In the midst of the hijacking crisis, a technofuturist on the U.S. West Coast, Sudarshan Khurana, a person who had spent a good part of his professional life working for a multinational corporation with operations in the Middle Eastern Islamic countries, as well as Russia, wondered whether a high-tech solution might prevent such a catastrophe in the future.[9] He suggested that the manufacturers of international passenger jets, the Airbus and the Boeing, develop and install a spray system that would pump in an invisible fainting gas, which, instead of inducing temporary blindness as pepper spray does, would quickly anesthetize the occupants. In a situation like the hijacked Indian plane, for example, pilot Devi Sharan would have simply pressed the panic button, and in seconds the nonsmelling gas would have put everyone, including the hijackers, to sleep. This technological solution, Khurana suggested, could work if combined with other technologies now under development in the United States and other countries. The inside of the airplane, for example, if any extraordinary emotional upsurge took place during the flight, could become intelligent and sensitive to human touch, just like the polygraph. Used mostly as a lie detector, the polygraph is actually a device for measuring different pulse rates due to an unusually stressful mental and emotional situation, such as sudden fear. The polygraph-sensitive skin of the interior of the airplane, networked via satellites with ground stations, could remain in a state of constant communication, like a space shuttle, networked car on a highway, or smart home. While this might sound like a James Bond/*Star Trek* fantasy, it is an achievable and desirable technological goal to make the world a safer place, though not necessarily a freer one.[10]

The technology of the Emotional Mouse, a fantasy gadget developed
at IBM's Almaden Research Center, in San Jose, California, as a part of
"affective computing" is a step in the direction of making our surroundings responsive to our needs. The Emotional Mouse monitors the
user's skin temperature, heartbeat, and other physiological signs, including the skin's electrical conductivity, and feeds all of the information into the computer to let it determine the user's state of mind. If the
pilots and passengers of Air India, of the planes that hit the World Trade
Center, and of Flight 93 (whose passengers reportedly struggled with
the terrorists) had had this gadget on their seats or armrests, linked via
a satellite to the ground station, these hijacking stories might have
ended differently. The release of terrorists from the Indian jail had serious consequences for the United States. One of them, Ahmed Omar
Saeed Sheikh, was later involved in the kidnapping and murder of the
Wall Street Journal reporter Daniel Pearl. Terrorism, like commerce, is
global and uses the same transmission belt.

TECHNOFANTASTIC

Researchers at IBM and the Massachusetts Institute of Technology
working on "affective" computing technology have a modest purpose:
to make the computer warm and friendly to the human touch.[11] They
want the computer to capture the user's emotional state and to utilize its
digital brain to adapt itself to the needs and better the mood of the user.
For example, if the Emotional Mouse found that the user was sick, the
computer might suggest calling the spouse or pipe in soothing music.
Apply these seemingly crazy ideas to the airplane in distress—the networked, sentient flying computer—and how its polygraph-sensitive
skin might sense the danger, then, like Hal in the movie *2001: A Space
Odyssey*, disable the hijackers as the West Coast technocrat Khurana
suggested. It might seem technofantastic, but so was the Arthur Clarke's
fantasy of the geosynchronous satellite in 1945. Now that wearable
computer devices are becoming commonplace, it is not far-fetched to
think of the inside of a tower in New York as being sensitive, intelligent,
and responsive to the emotional and psychological plight of the occupants and able to act appropriately. Airplanes and buildings, like large
supercomputers, William J. Mitchell wrote in *e-topia*, "will continu-

ously suck in information from their interiors and surroundings, and they will construct and maintain complex, dynamic information overlays delivered through miniature devices worn by inhabitants, screens and speakers in the walls and ceilings, and projections on to enclosing surfaces." In the coming e-topia, buildings under attack, for example, will not "just sit there, but actually consider what they are supposed to be doing and choose their actions accordingly."[12] So will underground sea tunnels, subway trains, and automobiles.

CAVEAT

While the United States and other countries fight terrorism with superior technology and indifferent diplomacy, it is important to remember that excessive hysteria about terrorism can build into a dangerous anti-Arab/Muslim mind-set that will prevent people from distinguishing the militant from the innocent, stirring religious and ethnic hatred and eventually distorting diplomatic and commercial relations. This is what happened in the case of a Dubai-based company's proposal to manage six East Coast sea ports; the proposal was withdrawn after the U.S. Congress threatened to block it. When humans lose their judgment, computers with artificial intelligence capability can only aggravate the problem. We must never stop reminding ourselves that the United States is an open, multicultural, and multireligious society, much admired by the rest of the world, and there lies America's moral strength. Joseph Nye of Harvard calls its openness the nation's "soft power," the power to persuade the rest of the world that fighting terrorism is a collective responsibility.[13] Nonetheless, digital technology is indispensable to building a virtual environment that makes the world collaborative, transparent, and open for everyone to share the fruits of globalization; it is also a necessary condition for the free marketplace to work.

Self-aware Collaborative Environment

What the scientific community, for whom the Internet was originally developed as a protected private channel for collaborative research, needs now is the equivalent of the *Star Trek* holodeck, which would enable users to have real-time, virtual reality collaboration, even when they were thousands of miles apart, so that companies in Bangalore and Boston would have no difficulty in sharing knowledge and expertise via cyberspace. Highly specialized expertise and brainpower dispersed in different world centers must be pooled and interconnected to solve crucial problems in real time. For advanced scientific, medical, and technological research, at present the Internet as a global collaborative network is only in the initial stages of development. Improvements in Internet technology and multimedia capabilities, such as three-dimensional computer modeling of the human heart, are necessary so that a cardiac surgeon ready to operate on a patient in London, for example, can confer and work jointly on a holographic platform with a colleague at Massachusetts General Hospital in Boston in real time.

For those dreaming about holodeck virtual reality science fiction possibilities, the next warp in cyberspace is the development of Internet2, which at present is being used only for top-priority scientific and medical research. As happened with the Internet, eventually access roads to Internet2 will be built for businesses and the general public.[14]

In 2000, 112 U.S. research universities and the National Science Foundation (NSF), along with several high-tech companies, formed a consortium to develop a faster system of internetworking. In May 2002, the University Corporation for Advanced Internet Development, a nonprofit group established at the University of Michigan, took over the consortium's drawing board initiative to develop the next generation of Internet technologies. According to Bill Graves, chairman of the Internet2 Applications Task Force, the goal is "to create a virtual environment as realistic as possible so multiple people in multiple locations can collaborate in cyberspace."[15] Building these virtual environments, where people living thousands of miles apart can work together, presents one of the most exciting technological challenges. Today's phone lines, which carry e-mail and other popular Internet traffic, assign equal weight to all information packages based on the net neutrality principle, regardless of their importance and urgency; this has to change fundamentally in order to make traffic flow more efficiently.

Apart from increasing the bandwidth for three-dimensional audio and video transmission, Internet2's protocol will have special commands that enable it to prioritize information and avoid congestion when a clear channel is needed for special applications. Space scientists in Houston and Moscow, for instance, could solve the international space station's problems with the crew in a virtual environment. Software solution designers in Bangalore and Stanford might work together to solve problems simultaneously as they arise in real time.

The Internet2 uses a high-speed national network run by MCI Communication Corporation for the NSF to connect the nation's supercomputer centers. Participating universities are responsible for building their own high-speed infrastructures and connecting them to the MCI-NSF network, for which the NSF gives them grants. The entire project was a prelude to the Clinton administration's $300 million Next Generation Internet (NGI) initiative aimed at eventually linking all of the nation's universities and laboratories.

Today, 207 universities, in cooperation with government and industry, as well as some international collaborators, are leading Internet2, but eventually, when it expands to accommodate commercial interests and the general public, there will be congestion and a cry for Internet3. Digital civilization thrives on continuous reorganization and speed.

With the emergence of the NGI initiative, the Internet might split into several different categories. There may be a separate Internet for e-mail, the most widely used form of international communication today, that does not require high-speed transmission for everyone. There could be a special Internet for high-speed integrated voice, video, and data transmission to be used for virtual reality conferences and limited to members and other select groups. There could be a separate Internet for global e-commerce. Eventually, there could be a universal matrix of Internets, evolving and self-organizing under the forces of the free marketplace.

The United States is the land of free market capitalism, of Henry Fords and Bill Gateses, but despite the openness of the marketplace, great technological ventures like the Internet, the National Aeronautics and Space Administration, and the Manhattan Project require a heavy infusion of national resources and a clear vision, and federal government support is indispensable in exploring the frontiers of cyber-space-time. In the ultimate analysis, however, the future of the Internet in the United States will be determined by the free marketplace and national security. In China, Iran, and Saudi Arabia, the freedom of the Internet will be determined by the trajectory of the pendulum controlled by each country's bastion of core values. It will be interesting to see how much digital freedom their systems can tolerate without being torn apart.

Cyber-space-time Access in the Digital Age

Just as the Internet raises questions about the meaning of freedom, it is also challenging the concepts of space and time in the temporal realm— as Einstein did in transforming our thinking about space and time on a cosmic scale. Time indeed runs faster in the cyber age, and a restrictive job contract, such as forbidding an employee to take a job with a competitor for a year or so, may be deemed an unfair practice, restrictive of personal freedom.[16] Similarly, access to cyberspace has become constitutionally as important as access to physical space.

A few years ago, the blind in the United States demanded access to cyberspace, asserting that like Ved Mehta and Helen Keller, they too could work like sighted people if given access to public facilities, including virtual space. This was not a plea for compassion but rather an assertion of their right to be in cyberspace, and they sued an online service provider to fight against cyberdiscrimination.

On November 4, 1999, the National Federation of the Blind (NFB), a leading advocacy group with a membership of approximately fifty thousand, filed a lawsuit against America Online (AOL) for not providing the blind access to public accommodation as defined under the Americans with Disabilities Act.[17] The act, which was signed into law by President George H. W. Bush on July 26, 1990, (probably his greatest contribution to the advancement of human rights in the United States), requires that new constructions, including renovations and alterations, as well as existing buildings and facilities, be made accessible to the handicapped. The law also requires that public services like transportation and telecommunications facilities be modified so that the handicapped can use them. All of these entities, whether owned and operated publicly or privately, fall under the term *public accommodation* if they affect commercial activities. The public accommodation section of the disabilities act includes access to motels, bars, motion picture houses, auditoriums, grocery stores, banks, hospitals, zoos, nurseries, and golf courses, to name a few. If some of these services and activities are also available on the Internet, it sounds reasonable to extend the law from physical space to virtual space. The disabilities bill was debated and passed in the pre–cyber age, when public accommodation meant only physical structures and cyberspace was still emerging from the scientific imagination. But the convergence of technologies and the development of the World Wide Web have revolutionized the way we do business, gather knowledge, and live our daily lives. Denying anyone access to cyberspace violates the equal-opportunity provision of the Fourteenth Amendment to the U.S. Constitution, as well as the disabilities act.

The NFB, according to the lawsuit, has been "actively involved in promoting adaptive technology for the blind" to enable them to live and work independently in today's technology-driven society. The International Braille and Technology Center for the Blind in Baltimore, Maryland, for example, trains thousands of blind people in the use of adaptive technology. The center's computers include screen-access software that enables the blind to use e-mail, chat rooms, and instant messaging by converting the text and data into computer speech or refreshable Braille, which allows the blind to read the computer screen line by line.

The plaintiffs, among them some very highly educated and fully employed blind people, complained that AOL, a self-proclaimed world leader in "interactive services, Web brands, Internet technologies, and electronic commerce services," required users to run its proprietary software on the Windows or Macintosh platform. But AOL's software did not convert information into synthesized speech or refreshable Braille, and thus discriminated against the blind and denied them "public accommodation" according to the disabilities act. AOL had excluded the blind from such normal online services as news, sports, games, finance, shopping, health, travel, updated stock portfolios, and blocked access to websites wrong for kids, which was a clear, hence actionable, case of discrimination.

The lawsuit questioned whether physical space in the Internet age means only physical structures. If cyberspace is an extension of physical space, online service providers and businesses must be subject to the provisions of the Americans with Disabilities Act, and their virtual services must be made accessible to the blind as sidewalks and ballparks are. Whether buyers go to Barnes & Noble to buy a book personally or they use the bookstore's website, they negotiate the deal in real space-time, though space may be virtual or physical and time may be synchronous or asynchronous.

In a 1994 case filed under the Americans with Disabilities Act, *Carparts v. Automotive Wholesalers Association,* a three-judge panel at the U.S. Court of Appeals for the First Circuit in Boston considered the question of public accommodation and ruled that access extended to telephone and the mail.[18] To the American people, the Internet has become as vital to existence as clean water, fresh air, and universal telephone service. Those previously marginalized due to physical or social disabilities are demanding their rights to be in the cyberspace mainstream so that they can be productive. The freedom of cyberspace must be extended to all citizens so that they can live full lives and be equally productive in the free marketplace.

DIGITAL POTHOLES AND TOLLS

Letting the market regulate itself is the essence of free market capitalism. The digital marketplace must be free from anyone's control, though in the course of time, policy guidelines for unobtrusive oversight may become necessary to prevent predators, criminals, and terrorists from stalking innocent people.

The Clinton administration wisely decided not to intervene in the commercial activities of the Internet and to let market forces, at least for a time, regulate digital commerce, probably in much the same way that borderless Asian-European commerce in ancient times happened because of the self-regulating Silk Road. Close to the end of his term in the White House, President Bill Clinton decided that his administration would do nothing to undermine the capacity of emerging technologies to improve the lives of the American people through an expanded marketplace. Ordinarily, a person will not care whether he buys goods on the Internet or in a shopping mall so long as he gets a decent bargain without being cheated. But the fear of cybercrime goes beyond the threat to fair-trade practices in the digital marketplace. Every time a person trades electronically, he leaves personal information behind captured by cookies and other software technologies that track digital surfing. Even the most innocent digital surfing can be stored in a large database, then mined to reveal patterns of personal preferences that might expose the user to surveillance by unscrupulous predators and traders. Moreover, since the Internet is a network of networks connecting millions of computers and devices, data can be intercepted at a node anywhere in the world. A car in distress might broadcast distress signals, but the message could be picked up by anybody, and instead of a rescue squad, a thug or a rapist could take advantage of the person in need.

Calls for privacy protection and personal and medical data security through encryption have been loud and shrill. But the Clinton administration resisted the sale of stronger encryption software technology for data scrambling for fear that it might interfere with the ability of the law-enforcement authorities to break into the secret world of drug traffickers and terrorists, who have been increasingly using encrypted communications on the Internet. The European Union, on the other hand, has adopted tougher standards for consumer privacy protection on the Internet, which has posed a great challenge to American companies doing business there.

TAXATION

The Internet is also creating taxation headaches for thirty thousand city and local authorities in the United States that cannot figure out how to collect their share of the burgeoning electronic commerce. How should Vermont, for example, collect tax dues on a transaction made by a resident who uses the Internet to buy goods from an online company in New Hampshire (a no-sales-tax state) to be sent to his friend in Colorado?

Unable to do anything about the taxation problem shadowing surging e-commerce, Congress decided to do nothing except pass the Internet Tax Freedom Act (1998), imposing a three-year moratorium on further Internet taxes.[19] The act, however, established an advisory commission consisting of state and business leaders to develop a policy for Internet-commerce taxation.[20]

State and local sales taxes are no laughing matter because they account for 40 percent of a state's revenues. In 1998, when the Internet Tax Freedom Act was passed, e-commerce transactions were estimated at $43 billion in the United States; they grew to $172 billion in 2005 and are expected to grow to $329 billion in 2010, according to Forrester Research. This will cost the states billions of dollars in lost sales tax revenues if nothing is done.[21]

There is another peculiar twist to the problem, which will create further inequity between Internet haves and have-nots. The poor who buy their goods from shopping malls and corner stores will be paying sales taxes, while the rich with access to the Internet who buy their goods online will not—an example of trickle-down economics in reverse gear.

Not to trivialize the matter, however, Forrester estimated that of the $43 billion in digital commerce in 1998, $925 million accounted for on-line pornography sales revenue, which is expected to grow 30 to 40 percent per year. Would the state and local authorities have moral scruples about taxing this revenue stream, even if it enabled them to fund the building of better schools, prisons, roads, and housing for the homeless? We are not ready to face the moral and ethical dilemmas created by temptations in cyberspace.

Temptations in Cyberspace

The illusion of freedom and control increases manyfold when a person has access to both a mouse and a gun. The forty-nine-year-old day trader who hammered his wife and two children to death, killed nine day traders in two brokerage houses in Atlanta where he traded stocks on the Internet, then shot himself on July 29, 1999, had used both a mouse and a gun to take control of his miserable life and free himself from it. "I have come to hate this life," wrote Mark Barton.[22] But he would not go away until he had destroyed the people whom he thought sought his destruction. Barton displayed some of the worst tensions to which digital Americans have become prone: a difficult marriage; inability to cope with the burden of children, leading to neglect and abuse; and the bull market temptation to make a quick killing online while the boom lasts.

For the bold and the venturesome, the gold rush on the Internet begins with starting a portal or search engine, as the Yahoo! and Google developers did, for example, then going public and becoming one of the legends of Silicon Valley. The lesser breeds, day traders, for example, rush to the mouse to trade on the stock market either from their home computers or at Internet brokerage firms. A day trader holds stocks for a very short time and trades several times each day, making money on margins and quick turnovers. Instead of going to the stock market floor and yelling buy or sell, the day trader watches the computer screen like one who is demonically possessed. According to some estimates, at its frenzy in 1999 and 2000, there were forty-five hundred to five thousand full-time day traders, some of whom had given up their full-time jobs.[23] They averaged about thirty-five trades a day, watching stocks' ups and downs, buying low and selling high. Besides, there were some two hundred fifty thousand other day traders who used their home computers for limited cybertrading, averaging two trades per day, on the side.

One can start with as little as $5,000, but, as with casino gambling, the novice trader does not know when to stop and continues trading on borrowed money, always hoping for the upswing, until losses mount and facing family and friends seems worse than death. The illusion of freedom and control at the click of the mouse is so intense that one cannot stop to ponder over the long-term consequences of one's digital actions. That was the downhill slope that Barton took: he went from chemical salesman to failed day trader, with the burden of two failed marriages, children to raise, and nowhere to hide.

In a burst of euphoria in the days of the dot-com boom, Andrew Shapiro wrote in *The Control Revolution* that the Internet gave people the means to control some aspects of life that had been hitherto under institutional controls, be they of the government, corporations, or the news media.[24] The truth is that, today, we have no control over ourselves or our work or the information we receive. In fact, some of us do not know how to handle the amount of freedom unleashed by the Internet and would rather live with rules and regulations imposed by the government or some other organization. Throughout this book the question reverberates, How much freedom does a person need? Indeed, after the slaughter at the Atlanta brokerage firms, the board of the National Association of Securities Dealers proposed rules obligating brokerage firms to warn day traders about the possibility of steep losses and the dangers of trading on margins and borrowed funds. Cyberspace, too, needs the rule of law to stabilize the system.

The mouse gives us unlimited access to information and unbridled freedom to play with bits and bytes, but no control over what happens once the bits and bytes submerge into the massive, invisible data stream swirling around the world, then finally coming to reside in a database where it can be used for purposes for which it was not intended. The postmodern digital revolution is both liberating and constraining us in the crucible of creative destruction.[25] It is giving us new power to click, create, and fake a myriad of cultural forms, which raise ethical questions as they expand the marketplace.

Click, Create, and Fake

Photographs are supposed to be inherently accurate and revelatory because they communicate actuality. The old adage, attributed to Nathaniel Hawthorne in *The House of the Seven Gables*, "A photograph does not lie," is no longer true because today digital technology can manipulate and change a photograph. It can remove an undesirable person standing next to you, place you in Buckingham Palace with the queen, or bring a long-lost cousin into the family-reunion picture. While this may all sound like innocent, creative fun, digital photo manipulation is raising serious questions about media creativity and credibility. Credibility rests on durable structures, but when structures keep moving, confidence is shaken. The fluidity of structures is a major characteristic of the digital civilization.

Before O. J. Simpson, a former national sports icon, went on trial for the double murder of his ex-wife and her boyfriend, *Time* used a digitally darkened mug shot on its cover, conveying guilt before due process had begun, while *Newsweek*'s cover presentation was neutral. The contrast between the covers of the two magazines raised a serious question about the rush to use technology without considering the individual's right to a fair trial based on the assumption of innocence, demonstrating the dynamic tension between the creativity of the free marketplace and the Sixth Amendment of the U.S. Constitution.

National Geographic is a trusted magazine, highly regarded for its superb photography and in-depth coverage of internationally important subjects, but its digitally moving one of Egypt's great pyramids to suit the design of a 1982 cover raised questions about whether reality should be altered for aesthetic reasons, when faking is a matter of few clicks. Where does reality stop and artistic manipulation begin? The marketplace demands complete disclosure, though some might call it professional ethics.

As the reader's surrogate eye, news photographers have an obligation to capture the moment accurately, thereby enhancing the trust readers repose in the news media. Trust is also a pillar of the free marketplace. A grocery store tabloid might titillate its readers with the picture of a two-headed goat or a woman's sexual encounter with an alien, but if a news organization like the *New York Times* or Associated Press were to use composite pictures without labeling them as such, little of their credibility would be left. Credibility is the backbone of a great news organization and more so in the digital age when factoids are pushing out facts. Credibility is also a valuable and indispensable marketing tool.

CREDIBILITY IN THE DIGITAL AGE

It is a tall claim that newspapers pursue truth; actually, they pursue stories. Presentation of facts accurately is, nonetheless, the beginning of any serious inquiry into truth. An altered photograph, regardless of its aesthetic appeal, is a lie and damages the integrity of the newspaper that prints it. But with digital scanners and desktop publishing becoming user-friendly and affordable, doctoring photos has become so easy that one is tempted to surrender to the Nike-popularized impulse and "just do it." Most people cannot discern the difference between an electronically manipulated picture and the original, which makes tampering with actuality, mostly for short-term commercialism, so irresistible, when one has only to click—just click. The mouse induces disinhibition and a false sense of power.

The National Press Photographers Association affirmed in 1999 that "the guiding principle of our profession is accuracy. . . . [A]ltering the editorial content of a photograph, in any degree, is a breach of the ethical standards."[26] But consider, hypothetically, the ethical dilemma of a newspaper that digitally manipulates the photo of a boy with an open zipper taken at a ballpark. If the newspaper uses the picture with the pants unzipped, his parents might file a civil suit against the newspaper for public embarrassment, even though the picture was taken legitimately in a public place. Instead of digitally altering the picture to close the zipper, however dramatic the picture, the photo editor might have asked a simple question: Can we do without this picture? No picture is totally indispensable to a story, but when the technology is available, editors forget their code of ethics, though they should not forget the threat of a lawsuit. It is important to distinguish between documentary photographs, altering whose content must be deemed unacceptable, and photo illustrations, which must be clearly labeled. The cutline could provide sufficient explanation of the digital special effects used in the development of an illustration, especially when the photo might lead to misunderstanding on the part of the viewer.

Some photographers, especially those working for advertising agencies and glossy magazines, argue that photo modification (retouching, for instance) has been practiced since the beginning of photography in the 1850s. Now that digital technology has enhanced their freedom, they can do much more with a photograph and exploit its full artistic and commercial potential. But they pay a high price for this transgression, namely, they lose their credibility as a reliable source of information if the intent of the photographer is to convey accurate information. In time, a legal distinction may develop between a photograph and its digitally altered version, the "computograph," and news organizations may have to vouchsafe photos' authenticity. Seeking credibility in cyberspace could become one of our greatest obsessions, as we seek digital synergy through interdependency and collaborative brain and muscle sharing.

Synergy through Global Interdependency

Richard Rosecrance argues in *The Rise of the Virtual State: Wealth and Power in the Coming Century* that in the twenty-first century, capital, information, and highly trained labor will make the old factors of land and political domination unnecessary. High-level manufacturing and services will become the most remunerative factors for the creation of the wealth of nations. Products of the mind, not land, will be the goals of "mind nations" as opposed to "body nations."[27]

Body nations, such as China and India, will manufacture goods for mind nations like the United States and European Union countries, while the mind nations will specialize in the creation of knowledge through research and development, branding, product design, marketing, patenting, financing, insurance, global transport (e.g., FedEx and UPS), and other mentally manipulative systems. Some countries like the United States will be both mind and body nations, though the manufacturing share of the U.S. gross domestic product will continue to decline.

Since in the digital age, dynamic systems keep changing and growing, body nations will not remain content with their inferior status as cheap labor markets. South Korea began as a manufacturing center but has been evolving into a mind nation. Its recovery from the 1997–1999 financial crisis was rapid and remarkable. It is quite possible that what has kept North and South Korean dialogue going is South Korea's realization that it needs to develop a body nation of its own rather than investing in China to take advantage of its cheap labor. Yet, think about what happened to Indonesia, once a manufacturing back alley of the

U.S. marketplace, which became, after the financial crisis, a hotbed of Islamic fundamentalism. It is not always possible to predict what direction the development of a society will take, unless we look at its bastion of core values, the invisible force that keeps it together and moves the cultural pendulum that describes its freedoms.

While talking of the synergy between knowledge-producing, product-designing mind nations and body nations that manufacture the products of the mind, Rosecrance hastens to allay our fears that this emerging relationship is not a newer form of imperialism. While we may not worry about another round of Opium Wars or the East India Company's rising on the bank of the Hoogly, nonetheless, we cannot forget how global energy companies try to bulldoze their way into dominance using their political connections in the White House. Nor should we forget how the United States abandoned Indonesia, once a darling body nation manufacturing cheap goods for the American consumer. As foreign capital began to flow out of Indonesia, all that was left was political chaos, heartbreak, and religious extremism. No one knew how to put together the disintegrating body nation, which in earlier times might have fallen prey to communism; now the danger is jihadist terrorism. In some countries, the pendulum of freedom moves between the marketplace and religious fundamentalism, as in Iran; in some others, it swings between nationalism and mercantilism, as in China. The drums of freedom beat differently.

MIND AND BODY NATIONS

Though China today is the world's leading manufacturing nation, its authoritarian rulers know how to extract concessions from nations with mind power, such as Japan, Germany, and the United States. China's lure is also due to hype about its billion-plus growing market, which hypnotizes European and American businesses. The well-scripted public relations discourse in the digital age has been helping China to attract billions of dollars of foreign direct investment and grow at a double-digit rate during the last two decades. Nationalist mercantilism predicated upon the global marketplace might have turned communism into a carcass in China, but it has not enhanced freedom, as Americans understand the meaning of the word.

What is transforming the world economy is the endless prospect of high-quality, cheap production in developing countries without the need to colonize and control them directly. This new global interdependency between body and mind nations may eliminate world conflicts in some regions, though one might wonder how long this arrangement will restrain China from forcibly taking over Taiwan, a country that has become a mind nation on its own and has been pouring billions of dollars in foreign direct investment into mainland China. Ironically, a body nation is threatening to incorporate a mind nation to gain a mind of its own or increase its mind power. When China's oil company CNOOK bid for Unocal, America reacted with fear and consternation that a body nation might take control of a company belonging to one of the most advanced mind nations.

Since the Internet connects everyone globally, it is possible for a factory in Ohio to move to Shanghai and still be managed by its U.S. owners without their setting a foot in the country. The reason American corporations like to invest in China is simple enough. China has not allowed the political conditions, dissidence, and other kinds of democratic noise generated by an open society to stand in the way of foreign investment. Corporate America does not care for democracy in China or the plight of poor peasants who have been driven off their meager land possessions to make way for highways and factories. Corporate America wants rules of engagement that favor its growth and profits. Efficient political management and a rule of law that protects foreign investment are the most important conditions for a mind nation to work with a body nation. China meets those conditions and is now rising to become a mind nation on its own, without freedom and democracy.

Rosecrance, however, makes a very important point, that mind nations cannot maintain their supremacy—design over manufacturing—for long unless they continue to upgrade their education and training, the ultimate determinants of what gives a nation mind power; Tom Friedman further developed this idea in *The World Is Flat*. Even in a flat world, however, a rich nation like the United States can import brainpower. Silicon Valley, for example, keeps clamoring for information workers from other countries and is urging Congress to raise the immigration quota for highly educated knowledge workers, particularly from India and China, who are needed to create digital synergy; at the same time, the rest of the country keeps protesting against illegal immigrants from Mexico, a country whose body workers are desperately needed in the United States.

DIGITAL SYNERGY

Synergy characterizes the ethos of the age of the Internet; it occurs when, through technological convergence, discrete businesses cooperate and combine creatively so that the emergent effect is far greater than the sum of the effects of the independent units separately.[28] To remain globally competitive, corporations seek synergy and self-renewal through mergers and technological convergence. Synergy can also be achieved by combining two or more systems into one modular whole, while maintaining their separate identities. Corporations desperately seek a functional model of synergy, and since no business school has risen to the challenge of providing one, synergy is left to the creativity of the free marketplace.

The search for synergy is not limited to the business world. It is equally important in nature. Ants, for example, build an enviable synergetic environment for survival. Ants transcend their smallness through synergy. The history of technology bears ample evidence that major technological leaps occur only through synergy. Samuel Morse's telegraph, for example, combined with Heinrich Hertz's electromagnetic spectrum theory to give birth to Guglielmo Marconi's wireless telegraphy, or radio, and subsequently to radio with pictures, or television. At every step of technological evolution, the combination of two or more discrete concepts or technologies has created synergism. For example, when the functionalities of television and computer are integrated into a cell phone, we have something new for which we do not have a name, though it is transforming communications as never before. The commercial hype of high-definition digital television is that it provides many different features simultaneously, for instance, picture-in-picture, so you can watch television and surf the Web at the same time. You can watch *Survivor* and participate in the show as well. You can watch football, chat with a friend, send e-mail, and read about the latest sports scandals on the same screen. But the cell phone, a marvelous example of converging technologies seeking synergy, is the future of communication.

Media companies, too, have been trying to better their chances for survival in the global market through synergetic convergence. NBC and Microsoft, along with *Newsweek* and the *Washington Post*, joined hands to create MSNBC.com, which sought to create an information-rich multimedia environment. ABC, CBS, the *New York Times*, the *Washington Post*, and other newspapers are being transformed into multimedia delivery systems.

Not to fall behind in the global race for synergy, Rupert Murdoch, whose satellite television reaches millions of homes globally, has realized that there is no future without the Internet, so he bought MySpace, a social networking website used mostly by the N-Gen (i.e., the Net Generation, which grew up with computers at home and in schools). He is raising a new corporation to link his satellite empire with the Internet to create not only an alternative delivery system but an always-on interactive communication world.[29] Today, viewers can shop, play games, buy corporate stocks and bonds, go to school, and do other hitherto unimaginable things in a virtual world of interactivity, for example, using Google, YouTube, Joost, and Second Life, which will require further reinventing of the Internet through the synergy of the TerraGrid.

SYNERGY THROUGH TERRAGRID

The Internet developed like a series of interconnected urban clusters, until the sprawl enveloped us globally. It was started initially in the 1980s for scientists by the NSF through its network NSFnet to create linkages for information sharing based on common transaction rules, or protocols. For most of the people outside academia and the research community, however, the Internet remained an esoteric enterprise until the World Wide Web democratized it in 1995, at which time the concept of hypertext was turned into a killer application by Tim Berners-Lee of CERN Laboratories.[30]

As the Internet developed and expanded into a medium of mass communication, peer-to-peer sharing, social networking, and business transactions, it became a convergence hub for multimedia applications, with the potential to create tremendous synergy. While increased use of the Internet created the problem of limited bandwidth capacity, it also posed the challenge of exploiting the unutilized capacity of millions of computers that lie dumb when not in use. When professors, for example, turn off their computers and go home, the university cannot use the computing power locked in on their desktops. Nor can the university rent that computing power out, particularly during long vacations in the summer when the student population on campus is very thin. If we multiply this unutilized computing power by the millions of computers in the country, we get some sense of the level of wasted resources. But this need not be so because the technology for using a slumbering computer's power is entering the marketplace.

The NSF undertook an ambitious project, the TerraGrid, by clustering four geographically dispersed supercomputing centers into one integrated network to make them work like one giant virtual computer whose power could be accessed from any of the four gateways.[31] The supercomputing cluster initially networked the San Diego Supercomputing Center at the University of California, San Diego; the California Institute of Technology, Pasadena, California; the National Center for Supercomputing Applications at the University of Illinois, Urbana-Champaign; and the U.S. Department of Energy's Argonne National Laboratory. The supernetwork clustering through the TerraGrid was expected to create eight times the power of the most powerful supercomputer today, capable of carrying out 13.6 trillion operations per second.

SUPERCOMPUTING BRAIN

The concept behind grid computing is enabling geographically dispersed computer resources to bundle and muscle together, coalesce as one brain, and provide the user with steady, consistent, and economical access to enhanced resources from any point of access. This newly unleashed supercomputing power, emanating as if from a single unified virtual source, could be used for solving large-scale data-intensive science applications, such as designing new drug remedies based on molecular modeling, climate modeling, weather forecasting, and modeling military operations. For example, in a grid-based virtual laboratory, a scientist can examine millions of molecules in the chemical data bank to identify and select those that have the most potential use for designing a new drug for breast cancer. Such a laboratory could be used for teleimmersion, giving users high-bandwidth access to a virtual environment, for example, a simulated surgical operation for medical students, an outbreak of bird flu for epidemiologists, or a battlefield for armed forces. It could be used for distributed supercomputing to solve problems that are too complex for a single computer.

But as the grid-computing technology leaves research labs and spreads to the outside world of commerce and trade, its consequences for business and education will be tremendous. Eventually all computers will be linked with local grids that will become part of the Terra-Grid, giving and receiving computing power and sharing databases and applications that may reside anywhere. Just as the TCP/IP protocol broke the system barriers and made the Internet possible, grid computing requires a single worldwide standard, and most bets are on the Globus, an Open Source system that will make it possible to tap the Internet on demand from anywhere. The Globus Toolkit, a collection of software applications and resources to support grid computing, can search where a particular database is located, decide how to divide and distribute a given amount of computing work among several computers on the grid system, and determine whether a user is authorized. Once a cell phone, desktop, or any other device is plugged into the global grid, not only it will draw its computational power from the grid but it can also use myriad applications, such as audio-video streams, databases, and videoconferencing, for example, simply as Web services, which will be available as utilities in cyberspace.

VIRTUALIZED POWER

Just as an individual household does not need its own power-generating system, a person needing twenty-five terabytes of computing power would not need his own supercomputer. He could simply tap into the grid to perform a specific task, and once the job was done, the resource would be returned to the grid for others to use. And those with excess computing capacity would be able to pool their computing power into a grid. Universities could form their own grids and be networked into national and international grids, creating the possibility of on-demand grid power. This is basically the concept behind distributed computing so that a user can remotely log into a grid system for one purpose and into another for some other purpose. The central processing unit need not be on every desk but could be virtualized for the benefit of everyone who could afford it.

The wireless revolution that's under way now will make grid computing a global phenomenon. Echoing William J. Mitchell's *e-topia*, Larry Smarr, an Internet pioneer, said, "Because of the miniaturization of components, we will have billions of endpoints that are sensors, actuators and embedded processors. They'll be in everything, monitoring stress in bridges, monitoring the environment—ultimately, they will be in our bodies, monitoring our hearts."[32] The first principle of the grid infrastructure has to be security. For example, the Globus enables each site in a grid to keep control over its resources, and only properly credentialed users can log in for job submissions. According to Irving Wladawsky-Berger, IBM's vice president of server group technology and strategy, "Grid Computing is the natural evolution of the Internet. . . . We are now entering the stage where access to applications, access to data, access to computing power, access to storage is becoming standardized through efforts like Web services and Grid Computing, which are defining those standards, and I believe that having these commonly agreed upon standards is going to make it easier for everybody to answer questions like 'do I want to do it myself, or simply access it as utility?'"[33]

Grid computing is one of those emergent technologies whose synergy is perceived to be so valuable that its universal acceptance seems inevitable if we want to develop global digital intelligence to expand the global marketplace.

Digital Intelligence Rising

In 1969, when the U.S. Department of Defense started the first computer network, ARPAnet, for sharing information among research labs, no one could have predicted that in less than three decades it would give rise to a digital revolution, one that is now sweeping the world and challenging societies based on absolute principles. There are technology naysayers who would like us to stop doing further research lest the future get out of control, particularly in the field of artificial intelligence, fearing that AI-embedded self-breeding and self-evolving software programs and machines might take over the Earth. It is possible that as chips and machines get faster, they might get smarter and also become capable of self-replication. Quickness of thinking is a function of intelligence; therefore, the more intelligent a person, the quicker is his thought process. A genius, for example, makes quicker and better connections to solve complex problems. A dull person thinks slowly.

Dr. Jordan B. Pollack and Dr. Hod Lipson, scientists at Brandeis University, have indeed shown the possibilities of a rudimentary form of self-replicating, self-evolving machine.[34] Pollack and Lipson provided their computer with a repertory of possible designs to choose from and instructed it to design a robot that would move according to the laws of gravity and friction. After hundreds of trials and errors, permutations, combinations, and simulations, the computer, without any further intervention from the scientists, created a model that was used to develop a prototype robot, which had enough intelligence to make creeping, crawling movements on its own.

Instead of having to build a prototype from the computer-created design, Pollack and Lipson were able to leave the digital object alone to evolve and self-replicate into thousands of complex designs. From this primordial cybersoup of designs, more sophisticated models could evolve and develop into cybercreatures with superior artificial intelligence. Something like that is supposed to have happened in the evolution of galaxies and life on Earth.

Now that the threshold has been crossed and the genie is out of the bottle, what will the consequences be? What might these self-evolving cybercreatures do? Will they undertake hazardous assignments like cleaning up nuclear waste, seeking terrorists, going into space, or fighting our dirty wars? And if they develop self-preserving survival instincts—instinct is a form of knowledge—then what will they do to us, especially if they become smarter than humans? Although not all technologies become acceptable in the marketplace, self-replicating and self-evolving machines do raise questions as serious as those surrounding the cloning of Dolly. Should research into self-replicating artificial life be limited or banned? Or should it be controlled by a few nations, as is the case with nuclear technology? It is possible that learning from each other's experiences in cyberspace, the self-reproducing and self-evolving robots might one day raise better robots than themselves or better even than humans. Perhaps our fears are groundless. Human beings, despite all their attempts at selective breeding and eugenics, have not bettered themselves in any significant way during the last two millennia, though they have become more lethal killers. Maybe the next step in evolution is human-machine synthesis, combining the best in computers, their speed and multiplexing abilities, with human consciousness, emotions, judgment, and ethics. But one cannot stop wondering with Bill Joy of Sun Microsystems whether this kind of research will aggravate the problem of "knowledge-enabled mass destruction."[35]

Brick-and-mortar civilizations depend upon physical structures, such as roads, highways, ports, factories, places of worship, books, and other artifacts, material things that tie the laws of men and gods to physical space, giving them the appearance of permanence and absoluteness. Postmodern digital civilization is based on dynamic databases, which are potentially recombinant and open to continuous reinforcement and upgrading. In comparison with the immobility of physical structures, database structures are fluid, interactive and dissipative, and self-renewing. Although they make rapid innovations possible, their fluidity and transportability create dynamic instabilities and insecurities, which challenge the traditional ideas of privacy, freedom, creativity, and control, as we are going to explore in the rest of this book.

Chapter Three

Changing View of Privacy

Most people grumble and protest about how the government and businesses are invading their privacy, and though they are not wrong in doing so, no one has come up with an idea of how much privacy a person needs and for what purpose. Like wealth, a person cannot have enough privacy, but not everyone needs the same amount.

For more than a century, privacy has been a matter of grave concern because many new, unobtrusive tools for invading privacy are continuously being developed.[1] George Orwell's novel *1984* frequently crops up in social discourse and leaves us pessimistic about the future. Big Brother has become a diffused digital omnipresence. Everywhere there are eyes and ears, though they may not necessarily be hostile.

The government has an enormous interest in gathering information about individuals for controlling crime and terrorism, to say the least. In a market economy that depends so much on consumer spending, it is a matter of survival for businesses to gather as much information about consumers as possible so that they can stay innovative and competitive. With the development of computer software that provides tailored business solutions, even a small firm can turn information into a marketing tool. If businesses are totally prevented from gathering information about consumers, the economic consequences will be catastrophic. Reliable information is necessary both for good government and efficient business. The question that arises is not so much whether the information gathered will be used for benign, legitimate purposes but whether its misuse can be prevented and how. Consider a person who has a health condition, cardiac arrhythmia, for example, and uses an online pharmacy to order the prescription drug prescribed by the physician. Should the pharmacy be prevented from sharing the information with health and auto insurance companies, which might raise the person's rates? Or do these companies have the right to gather information so that they can do business more efficiently? At the heart of the insurance business is risk assessment, and an insurance company should have the right to access complete and accurate information about the people it insures so that proper rates can be calculated. The state motor vehicle department has a legitimate interest in knowing whether a person might have a heart attack while driving, as does the community in which the person lives. To assert, in the name of privacy, that cardiac arrhythmia or any other critical medical condition is the personal affair of the individual and that no one else should know about it is socially irresponsible. The need for privacy, which includes the freedom to be left alone among its dimensions, is predicated upon what the marketplace can bear in the United States.

Privacy is legitimate if it causes no collateral damage to a society and its core values. And since core values differ from society to society, the legitimate expectations of privacy vary. Nonetheless, concerns about a technology-driven invasion of privacy cannot be dismissed out of hand because the very technology that makes life comfortable in some ways may chain us to invisible bondage or expose us to public ridicule.[2] Invisible bondage could take the form of continuous blackmail, for whatever purpose, if there is no recourse to a speedy and fair justice system. But no less reprehensible is the public ridicule to which we may be subjected when privacy is breached. Louis D. Brandeis, a highly regarded Supreme Court justice whose views on privacy have influenced modern thinking about the issue, wrote that the founding fathers "sought to protect Americans in their beliefs, their thoughts, their emotions and their sensations. They conferred, as against the Government, the right to be let alone—the most comprehensive of rights and the right most valued by civilized man."[3]

This indubitably aristocratic view of privacy found fuller discussion in the article Brandeis wrote with his law partner, Samuel D. Warren, in 1890 in the *Harvard Law Review*. The two well-to-do lawyers from Boston, before they went on to become Supreme Court justices, bemoaned the intrusiveness of the yellow press, which had begun to use the newly developed miniature cameras that enabled photojournalists to shoot without the knowledge or permission of their subjects:

> The press is overstepping in every direction the obvious bounds of propriety and decency. Gossip is no longer the resource of the idle and of the vicious, but has become a trade which is pursued with industry as well as effrontery. To satisfy a prurient taste the details of sexual relations are spread broadcast in the columns of daily papers. To occupy the indolent, column upon column is filled with idle gossip, which can be procured by intrusion on the domestic circle. The intensity and complexity of life, attendant upon advancing civilization, have rendered necessary some retreat from the world, and man, under the refining influence of culture, has become more sensitive to publicity, so that solitude and privacy have become more essential to the individual; but modern enterprise and invention have, through invasion upon his privacy, subjected him to mental pain and distress far greater than could be inflicted by mere bodily injury.[4]

SHARED LIVES

The framers of the Constitution did not share this hyperbolic view of privacy. Nor did they include privacy explicitly in the Constitution because it was not threatened the way it is now. The American people, during the time of the War of Independence, lived shared lives and did not have the same idea of privacy as they have now. Most people in eighteenth-century America lived on farms and in rural areas, where homes were simple structures that included all functions—eating, sleeping, and resting—on one level.[5] America was an oral society, where people shared both space and information with one another without fear of repercussion or consequences they could not manage themselves. In fact, they were discouraged from living solitary, isolated lives. According to Robert Smith of *Privacy Journal*, "[T]he church regarded living together in a family unit as the best possible arrangement," particularly for the young and for widows.[6]

The Fourth Amendment, which protects individuals from unreasonable searches and seizures by the government, along with the Fourteenth Amendment's due process clause, has been interpreted to further strengthen Americans' right to be left alone in their homes. People living in eighteenth- and early-nineteenth-century America had very little

need for privacy and were unself-conscious about it. Nor did they regard privacy as a necessary condition for living a good life. According
to Justice Raymond E. Peters of the California Supreme Court,

> In many respects a person had less privacy in the small community of the
> 18th century than he did in the urbanizing late 19th century or does today
> in the modern metropolis. Extended family networks, primary group rela-
> tionships, and rigid communal mores served to expose an individual's
> every deviation from the norm and straightjacket him in a vise of back-
> yard gossip, which threatened to deprive him of the right of "scratching
> where it itches."[7]

Although privacy concerns began to rise with the advance of communications technologies, particularly with the development of photography and the portable miniature camera in the nineteenth century, the
history of privacy in the United States does not tell of its progressive
loss over time; rather, it tells of the rising awareness of privacy as something worthwhile in itself, probably even a necessary condition for the
pursuit of happiness, as Brandeis and Warren thought. Most of all, it has
been recognized as something that can be transformed into a materially
valuable asset for the marketplace.

Limits on privacy were the price one paid for living in society in the
past, and even today most of us gladly give up some privacy to share
our lives with others and reap social benefits, without which the pursuit
of life, liberty, and happiness is not possible. Consider the case of a
pregnant woman who has no choice but to expose herself to the gaze
and hands of her gynecologist; the presence of an unauthorized person
in the delivery room, however, could be considered an invasion of privacy.[8] If, on the other hand, the child's delivery is presented as a performance and videotaped, and the woman is properly compensated, her
privacy concerns might diminish. Privacy has an exchange value, and if
the price is right, many people will sell their privacy rights.[9]

Privacy is also a matter of who is watching you. Although a man's
home may be his castle, no one wants to live in isolation.[10] Those who
are privileged to live in modern mansions sometimes pathetically seek
both publicity and privacy; unfortunately, the two do not go together, as
we see in the case of Hollywood celebrities who need publicity to market themselves and their movies but seek privacy in their personal lives.
Whether it was William Pitt or, a century later, Louis Brandeis and

Samuel Warren who bewailed intrusion into their walled existences, their idea of privacy was primarily limited to being secure from prying eyes. A person's seclusion has been considered important, supposedly because it is in seclusion that a person can maintain his autonomy and personal integrity and live without shame or ridicule. On the other hand, major religions of the world advocate some form of communally shared and open life because a transparent lifestyle makes it easy for a person to surrender to the will of God or to the church, the cleric, or some other religious authority. The wage of privacy is sin, which in some religious traditions has to be purged through confession, self-exposure, and repentance, which is tantamount to giving up one's privacy. Some politicians too give up their privacy when they seek public office; for example, Newt Gingrich admitted in a two-part *Focus on the Family* radio interview with Dr. James Dobson on March 8, 2007, that he was having an extramarital affair during his time as the house speaker while he was busy pursuing President Bill Clinton's impeachment in the House. Privacy might become an unnecessary burden if Gingrich decides to declare his candidacy for president of the United States, although in the process of self-disclosure for political purposes he might have violated the privacy of his ex-wife and that of his present wife, who was his lover at that time.

In the United States, over time, the right to privacy expanded to include not only control over one's physical space (against intrusion) but also ownership of one's physical appearance, gestures, and voice for commercial exploitation. The right to privacy also includes protection from a false-light invasion of privacy; this occurs when people feel embarrassed by having actions they did not perform wrongfully attributed to them, as sometimes happens in docudramas. Even telling the truth about a person has been deemed an invasion of that person's privacy and has triggered an action for damages (read: monetary compensation).

Privacy today is a legal and commercial battlefield, where much is at stake. It has become a commodity in the marketplace, and there are people who would gladly exchange it for material benefits. Some people also use privacy as a strategic tool for doing things that they would not do if they lived the shared lives of their colonial ancestors. As society criminalizes certain activities, some people raise walls not so much for the sake of protecting their privacy but to keep the government away from their unauthorized and unlawful activities.

WHEN TECHNOLOGY ENHANCES POLICE POWER

It is not only dictatorial regimes that look for new technological means of suppressing their people. Free societies, too, use new communications technologies to control people in the name of law and order and fighting terrorism. Not long ago, the U.S. Supreme Court looked at thermal-imaging technology and whether its use by law-enforcement authorities violated a person's Fourth Amendment protection from unwarranted search and seizure, even though there is no physical intrusion into a person's home.

The scientific study of heat emanating from an object is called thermology, which is finding remarkable applications in diverse fields, such as crime detection, the diagnosis of human and animal diseases, and firefighting. Heat emissions vary from one object to another. Human beings, animals, trees, and even buildings emit different intensities of heat radiation, which can be evaluated and measured with an infrared thermal-imaging camera to identify the source of the heat and of the problem. The thermal camera turns the emissions into digitally processed images, showing what might be happening inside the source. A change in the blood flow to a person's brain, for example, would show a thermal response that could be measured in color gradation with the help of a thermal-imaging camera.[11] The technology can be used to diagnose circulatory and nerve injuries in patients suffering from migraine and toothache, for example. Thermal imaging is also being applied as a tool for cattle health management, where it is "used as an estrogen detection system to find animals in heat, test bulls for fertility or check horses suffering from lameness."[12] Firefighters use thermal-imaging cameras to find their way safely through a burning building. Let us keep in mind that every new technology creates a social dialogue not only about its long-term effects but also about what else can be done with it. Particularly, law-enforcement authorities look to adopt new technologies to their own purposes. So do criminals. Ironically, by defending the fundamental rights of criminals under the Constitution, freedom has been enhanced for everybody, as the following case shows.

THE KYLLO CASE

In 1992, Danny Kyllo was charged with growing marijuana in the attic above his garage in Florence, Oregon. Though a man's home may be his castle, he cannot seal himself in and do whatever he wants because individual freedom must be balanced with the larger interests of society. A person, for example, cannot claim privacy under the Fourth Amendment and make bombs or grow marijuana at home without being held accountable when caught. At the same time, the police cannot enter anyone's house without a search warrant because the Fourth Amendment protects individuals' homes. The police have to show probable cause to the court that some specific law has been violated before they can enter the premises or put a wiretap on a person's phone or other telecommunications device.

In the case of Kyllo, based on intelligence gathered from neighbors, the police directed an infrared thermal-imaging device, Agema 210, at his home to measure the unusual heat pattern, the kind of heat generated by artificial light used to grow plants inside a home, which was radiating from the roof and wall of the garage. Based on the thermal-imaging information, the police obtained a search warrant and found that Kyllo was indeed growing marijuana plants, and he was arrested. The U.S. Supreme Court took up Kyllo's case to decide whether the use of thermal imaging by the police without a search warrant violated his Fourth Amendment rights, despite the fact that he had entered a conditional guilty plea at the trial court.

Unlike in a trial court, where the judge and the jury passively listen to the probing questioning of witnesses and experts, with occasional dramatics thrown in, the U.S. Supreme Court justices ask fundamental constitutional questions that are sometimes so profound, one is likely to get carried away by their rhetoric. They sound like the wisest people in the United States, trying to leave indelible marks on history and to influence the future of society. Sometimes they do. The U.S. Supreme Court has been a deep fount of philosophical thought in the United States and sets the future course of society in many ways.

The constitutional question before the justices in the Kyllo case was whether using a nonintrusive device like the thermal imager to read and interpret heat emanating from a house constitutes an unwarranted search. What if, instead of marijuana with halide "grow" lamp, a person were using a sauna bath for three to four hours? wondered Justice Stephen Breyer. In such a situation, how the police interpreted the heat emanating from the house could be very embarrassing. Heat emanating from a house is not the same as a person putting the garbage outside his home, because garbage disposal is a deliberate act; heat emanation is involuntary. The police can search the garbage without a search warrant because the person has relinquished control over it. But do the police have the right to record, from a safe distance, sound vibrations and heat emanating from a house for law-enforcement purposes without a search warrant as technically they do not enter the house? Is there "a reasonable expectation of privacy," and how far does it extend?

Modern technology is giving new weapons to the police to demolish our walls, enter our bathrooms and bedrooms, and do whatever they want in the name of law and order. Justice David H. Souter said, "I think there is a reasonable expectation of privacy that what you are doing in your bathroom is not going to be picked up when you take a bath with one of these thermal imaging [devices]."[13] The Bill of Rights may dissipate by pixels and bytes unless the people assert their rights, the Court seemed to be saying.

The police argue that the use of invasive technology is necessary to maintain law and order so that we can enjoy our civil liberties. They want to present a good image of themselves and want us to believe that they do no evil, even when they use brute force, eavesdrop on our conversations, tap our telephones, and use surveillance technology to invade our privacy. They assert that sometimes unavoidable circumstances compel them to physically restrain and even beat up violent and defiant suspects. When they are left with no choice but to shoot while pursuing criminals, in self-defense or to prevent greater harm, they are accused of going on a shooting spree. The melodramatic view of law enforcement is reinforced by popular American television shows such as *Law and Order*, *Texas Ranger*, and *CSI*, which feed our moral fantasies that evildoers do get punished, if not by God, well, then at least by the police.

TESTING THE RULE OF LAW

Bad people test our faith in the rule of law. Most Americans wanted Timothy McVeigh, the convicted killer of the innocent in the 1995 Oklahoma City bombing, to be executed without delay. When his execution was delayed due to some procedural lapses, there was anger and frustration with the slowness of the justice system. Our moral fantasies, the triumph of good over evil, began to clash with the most important principle of the rule of law: giving the benefit of the doubt to the accused, without which a civil society cannot exist. Speedy justice is the ideal, but justice delayed is better than justice poorly done. When in doubt, the criminal justice system in a civilized society will let ten criminals go free to save one innocent person. All you need is a single doubting Thomas among the jurors in a criminal case to acquit, especially when a heinous crime has been committed, which requires proof beyond a reasonable doubt.

Law-enforcement authorities are under great pressure to get convictions for serious crimes, partly because that is how their job efficiency is measured, but more so because the public becomes emotionalized by cases like those of O. J. Simpson, the sports celebrity acquitted for the double murder of his ex-wife and her boyfriend, and Timothy McVeigh, whose death by lethal injection drew worldwide attention. The police in such cases go out of their way to ensure that they get unimpeachable evidence, proof beyond any reasonable doubt. To stand up to intense media scrutiny and the aggressiveness of high-priced, amoral lawyers during a courtroom trial, the police must gather evidence only in the legitimate way, but they will do that in any way they can.

In the process of gathering evidence, the police sometimes run into insurmountable constitutional barriers, one of which is the inviolability of the home. In the times of the founding fathers, the Fourth Amendment was carved to prevent the police from physically trespassing into a person's home; the second half of the twentieth century, however, brought new surveillance technologies—telephone bugging devices, surveillance cameras, and heat-radiation measurement instruments— that enable the police to draw inferences about what is going on in a person's home without physically entering the constitutionally protected inner sanctum. When technology trumps fundamental rights, the Supreme Court steps in as the final authority to interpret the meaning of the U.S. Constitution, what the founding fathers wanted to say and what they would say if they were resurrected from the dead.

The Kyllo case assumed a large constitutional significance because, due to sophisticated modern surveillance technologies, from satellite imaging technology to online data sniffing devices like Carnivore, which can read our e-mails, our homes are being turned into glass houses, conjuring up a scary vision of the government's newfound capacity to intrude. That's how the U.S. Supreme Court concluded the

Kyllo case, and a 5–4 majority ruled on June 11, 2001, that the Fourth Amendment draws "a firm line at the entrance to the house"[14]; whatever happens inside a home is intimate and private. The government must be kept from prying into people's homes, now that technology makes warrantless prying so easy, because otherwise there is the danger that increased power might be misused.

The ruling in Kyllo's favor did not mean that the Supreme Court was sympathetic toward the illegal act of growing marijuana. It simply upheld the right of Americans to be left alone in their homes and be protected *from the excesses of the government*. If a person does something illegal, the means of collecting evidence against him by law-enforcement authorities must not be illegal. That is the foundation of a civil society based on the rule of law.

REASONABLE EXPECTATION

The Supreme Court majority led by Justice Antonin Scalia tried to explain under what circumstances a warrantless search is said to have occurred, when there has been no physical trespass into a person's home, thus violating the person's Fourth Amendment rights. In a 1986 case, *California v. Ciraolo*, the court made a precedent-setting ruling that the Fourth Amendment does not prevent law-enforcement authorities from walking with eyes open and visually observing the house of a person that is visible from a public place. Visual surveillance is no trespass, but the use of electronic listening devices placed on the outside of a telephone booth does violate a person's Fourth Amendment rights because the individual has a reasonable expectation of privacy in the telephone booth. Justice Scalia referred to *Katz v. United States* (1967), a case about telephone eavesdropping by the police, in which Justice John M. Harlan had enunciated a test: "[A] Fourth Amendment search occurs when the government violates a subjective expectation of privacy that society recognizes as reasonable." Applying the test, Justice Scalia said that the court had held in two cases that "it is not a search for the police to use a pen register at the phone company to determine what numbers were dialed in a private home, and we have applied the test on two different occasions in holding that aerial surveillance of private homes and surrounding areas does not constitute a search."[15]

A couple making love beside the swimming pool or growing marijuana in the backyard of their home cannot expect their privacy not to be violated by aerial police surveillance. In other words, all areas of a person's home are not equally protected from searches by the police. The problem nonetheless is that due to advanced technology, even the innermost sanctum of a person's home is not safe from the government's eyes. Echoing the increasing concern of the American people about the danger to personal privacy posed by such unobtrusive, intrusive technologies, Justice Scalia said,

> It would be foolish to contend that the degree of privacy secured to citizens by the Fourth Amendment has been entirely unaffected by the advance of technology. For example, as the cases we discussed above make it clear, technology enabling flight has enabled to human view (and hence, we have said, to official observation) uncovered portions of the house and its cartilage that once were private. The question we confront today is what limits there are upon this power of technology to shrink the realm of guaranteed privacy. . . . [I]n the case of the search of the interior of homes—the prototypical and hence most typically litigated area of protected privacy—there is ready criterion, with roots deep in the common law, of the minimal expectation of privacy that exists, and that is acknowledged to be reasonable. To withdraw protection of minimum expectation would be to permit police technology to erode the privacy guaranteed by the Fourth Amendment.[16]

The police's attempt to gather information, using sense-enhancing technology, from the interior of a home, where the expectation of privacy is greatest, violates the legitimate Fourth Amendment expectation of privacy, especially when such a technology is not commonly available to the public. In other words, the court majority seems to say that when thermal-imaging technology becomes as commonplace, for example, as telescopes, then, of course, the reasonable expectation of privacy might change. The court rejected the view of the government and the dissenting minority that the thermal-imaging device measured heat emanating "off the walls," over which the plaintiff had no control, that the device had not penetrated the interior of the house to capture intimate details of the goings-on within, and the police had simply inferred from the heat emanation that the plaintiff might be engaged in an unusual activity. "In the home," Justice Scalia said, "all details are intimate details, because the entire area is held safe from prying government eyes."[17]

The case was decided before September 11, 2001. Would the justices have thought differently if Kyllo were found making a bomb?

SNIFFING CRIMINALS ONLINE

Terrorists, drug traffickers, and organized crime groups, both domestic and foreign, cannot carry out their plans without access to telecommunications networks at some point. Since the criminal justice system demands evidence that cannot be discredited in court, electronic surveillance gives jurors unimpeachable evidence to determine facts based on a defendant's own communication. There is nothing more damning as courtroom evidence than e-mail and other forms of electronic transaction.[18]

Carnivore, developed by the FBI, was a clever surveillance technology used for apprehending and preventing crimes in the digital age. Like a giant fishing trawler, Carnivore sat across a data stream and caught all information passing through the portal of an Internet service provider (ISP). When attached to an ISP's server, it enabled the FBI to read e-mail subject headings and to record receivers' and senders' addresses. The technology was based on packet sniffing, a technology networks use for tracing lost e-mail files and for other maintenance functions. According to the FBI's official website, the FBI and ISP would determine an access point through which a suspect's information flowed and install a tap that copied all data at the access point.[19] The copied data was collected in a storage system, where it was filtered for the precise information that the court order had warranted. The FBI site assured that only relevant data would be collected and archived. "All information collected is maintained and, in the case of full content interception, is sealed under the order of the court . . . and may subsequently be made available by the court to the defendant."[20]

Despite all the constitutional checks, Carnivore gave law-enforcement authorities so much power over private data flow that many people became alarmed about its potential misuse and the violation of their constitutionally guaranteed right to freedom from unwarranted searches and seizures. The FBI said there was nothing to worry about because strong procedural safeguards were in place to prevent the technology's misuse. Under Title III, for example, a high-level officer of the Department of Justice must authorize the request for interception before the attorney general's office could apply for an order with a federal district court. Only certain specific federal felony offenses qualified for electronic surveillance and interception. The original purpose of electronic surveillance law was the gathering of evidence, not intelligence, but after the horrific events of September 11, 2001, to prevent further terrorism, the FBI has been given the authority under the Patriot Act to conduct intelligence gathering as well.

Traditionally, when authorized to use electronic surveillance based on the showing of probable cause, U.S. law-enforcement authorities have resorted to telephone tapping, for example, to eavesdrop on drug dealers (who use encrypted communications). Under the court order, telephone companies have to cooperate with authorities by handing over the telephone records of people under suspicion or letting them tap into these individuals' conversations. But such wiretapping is allowed only for people suspected of criminal activities.

Carnivore scanned all Internet traffic, including that of innocent people that passed through the portal of an ISP. Since it did not conform to the reasonable expectation of privacy standard, vide the *Katz* case, it was not only unconstitutional but it violated the confidentiality agreement between an ISP and its other customers, against whom there was no search warrant. Carnivore could not have done otherwise, though, because digital communication via the Internet is chunked into packets to be sent via different routes to their destination, where they are finally reassembled. Carnivore had to scan all headings in the data stream to select the ones it was programmed to pick up. In other words, we had to depend upon the honor of law-enforcement authorities that they would intercept only those communications authorized by the warrant. This amounted to putting too much trust in the FBI, for which there is no historical justification, looking at its past excesses.

Since most communication now passes through cyberspace, the government feels it is necessary to do cybersurveillance on its own citizens, especially after September 11, in order to maintain national security and protect critical infrastructure, including the information superhighway, without which the burgeoning e-commerce would suffer a setback. Secure communication is essential to economic growth. The apparent goal of Carnivore was to capture cybercriminals, some of whom stalk dot-com companies intent on blackmail and sabotage.

Every year, hackers break into financial service companies' systems and steal customers' home addresses and credit card numbers in order to commit identity thefts. Criminals have been using Internet communications to perpetrate all kinds of fraud on victims all over the world. Jihadist terrorists use the Internet for recruitment and fund-raising. FBI Lab Division Assistant Director Dr. Donald M. Kerr told a House of Representatives subcommittee, "The Carnivore device provides the FBI with a 'surgical' ability to intercept and collect the communications which are the subject of the lawful order while ignoring those communications which they are not authorized to intercept."[21] Carnivore met the stringent requirements of the federal wiretapping laws, he said, and was a necessary tool for law enforcement. Despite all the protestations of respect for individual rights, there is no doubt that Carnivore was a technology potentially made for Big Brother; therefore, the question was how to watch the FBI so that it did not exceed its authority, how to strike a balance between cyberprivacy and the needs of law-enforcement authorities.

There was a tentative proposal to update the present disparate electronic surveillance laws applicable to telephone, cable, and other technologies and to establish a uniform standard for all communications technologies, though nothing has resulted from such perfunctory efforts to streamline the system. Nonetheless, the court could not have stopped FBI's Carnivore from reading the e-mail messages of innocent people, along with those of targeted criminals, because the sniffing technology is nondiscriminatory at present. Carnivore might be more suitable for authoritarian regimes like those in Singapore and China and less suitable in the United States, where people are suspicious of the FBI. In March 2006, the FBI announced that it had retired the surveillance technology—but that did not spell the end of surveillance, as we shall discuss in the next chapter. Nor did it mean that the technology would not reappear under a different name or that other countries, where privacy is not as broadly defined and protected as in the United States, are not using it.

While the U.S. authorities are apologetic about surveillance technologies as a necessary evil to protect society from an even greater evil, British authorities have unabashedly gone ahead with establishing a $38.75 million Government Technical Assistance Center to catch cybercriminals under the Regulation of Investigatory Powers Act, which enables them to spy on their citizens' e-mail and encrypted communications without any prior court approval.[22] Regardless of whether there is probable cause, an ISP can be ordered to send data to the center at its own expense, and unlike in the United States, no scrutiny by a federal court is required. The act also gives the authorities the power to demand the encryption key to decrypt any intercepted communications, which no doubt compromises civil liberties in Britain and takes the mother of parliamentary democracy one step closer to resembling Singapore's excessively regulated society.

The American people, on the other hand, have always been more protective of their civil liberties and have a healthy distrust of the government. They feel that there is a tremendous potential for the FBI to abuse the technology on its own or under the prompting of the White House, as happened during the Nixon era. But the threat of terrorism looms large in the United States, and when American officials say that terrorists and hostile nations have extended the field of battle from physical space to cyberspace, against critical military systems and economic bases, most people fall into step with their drumbeat.[23]

DELTA FORCE TO NETFORCE

There has been speculation that not only al Qaeda but domestic terrorists must be training an army of super programming brains and computer wizards because future battles will also be fought in cyberspace. Consider a Hollywood scenario. Deputy Commander Alex Michaels of NetForce, a newly established division of the FBI that controls rampant organized crime on the Internet, has been pretty busy finding the murderers of his boss Steve Day, whom he replaced as the division head. In pursuit of the killers, Michaels discovers a plot by a megalomaniac, the CEO of a Silicon Valley company, Will Stiles, who walks and talks like a 007 villain and plans to break into the White House to steal the president's master computer code, which will give him access to any computer using his company's browser. All this happens while Stiles is entertaining the president, the first lady, and their entourage at a dinner at his palatial residence. The programming genius behind the scheme to take over the world via the World Wide Web is a dark-skinned, foreign-sounding Uday Shanker, presumably a person of Indian origin. This scenario is the science fiction brainchild of Tom Clancy, the best-selling author of technothrillers like *The Hunt for Red October* and *Patriot Games*. The two-part miniseries *NetForce*, broadcast on ABC, was only a virtual reality entertainment, but popular fiction and television melodramas embody our hopes and fears as they thrill us with their melodramatic escapades.[24] Programmers like Uday Shanker may threaten some Americans, but in the age of global interdependency, there is no choice but to hire the best brain available anywhere. In the crevices of global interdependency, however, terrorists can find safe hiding places, hence, the diffused paranoia.

But Tom Clancy is not the only one worried about terrorism, while at the same time cashing in on America's concerns about the cyberfuture. Michael Vatis, the chief of the FBI's National Infrastructure Protection Center, told the Senate Judiciary Subcommittee on Technology, Terrorism, and Government Information in 1999 that a "half-dozen substantial attacks" were attempted every day on government computer systems, including the Department of Defense's computers; his refusal to elaborate about the threat against what Senator Jon Kyl (R-AZ) called America's "soft digital underbelly" made it seem all the more menacing.[25]

It's difficult to say how much of this concern is panic driven and how much is realistic, but when compounded with the possibility of biological and chemical terrorism, it does make U.S. military operations, banking and finance, and power and transportation nodes—all dependent upon networked access points—look very vulnerable to domestic and international terrorism. The Clinton administration established the Critical Infrastructure Assurance Office to oversee the nation's telecom and information technology systems, including cable, satellite, paging, wireless, and emergency broadcasting, to meet the challenge of an "electronic Pearl Harbor."[26] Another Pearl Harbor? That's how Lt. Gen. Kenneth Minihan, the director of the National Security Agency, described the threat before the Senate Government Affairs Committee in 1999. The Japanese attack on Pearl Harbor on December 7, 1941, took America by surprise and was a turning point in World War II, as September 11, 2001, was for U.S. foreign policy.

Apart from the axis of evil mentioned by President George W. Bush, who are America's enemies? And where are they? They might be foreign national intelligence operatives, militant organizations, foreign industrial competitors trying to steal trade secrets, new-age terrorists, drug criminals, and unhappy, treacherous citizens. Or they might be part-time hackers and crackers for whom the greatest challenge is to break into the Pentagon computer system.

Think about the new-millennium plot of Tom Clancy's fictional Silicon Valley evil genius, Will Stiles, who wanted the whole world to use his Web browser (this is reminiscent of what Google and others probably want to do with information searches and storage) so that he could control the world via the Internet. Think about the possibility of a worldwide networked system's crashing or being hacked into if most of the computers are dependent on only one operating system. Add to this scenario the horrors of a centralized data system, and you have the makings of a digital Armageddon. Not only is decentralized digital power redundancy essential to fighting terrorism in cyberspace but it is also a pillar of digital postmodern civilization, watched over by NetForce, the equivalent of the Delta Force of earlier times. In this environment of digital terror, most people have been supportive of government efforts to fight terrorism, but privacy has been shrinking.

WIRELESS MOBILITY, FREEDOM, AND PRIVACY

Concerns about personal safety, 24/7 mobile business environments, and the need for social networking are pushing many people into going wireless. The *Herald* (Glasgow, Scotland) reported the remarkable story of the rescue of a fifty-nine-year-old injured climber from Cheshire who fell deep into the Llanberis Pass, Snowdonia, Scotland, crushing his chest and injuring his legs. Thanks to his presence of mind and the cell phone, he was able to reach the police, who, though they could not trace him, passed on the message to the Royal Air Force (RAF). An RAF Sea King helicopter rescuer also failed to locate the climber or to establish a contact with him on his cell phone. Then, a flight officer at the Kinloss RAF base had the ingenuity to send the lost climber a text message on his cell phone asking him to contact them, which he did, and they were able to locate his precise coordinates and rescue him.[27] A text message can get through if the signal is weak when a voice message may not. Wireless phones, along with 911 free calls to the police, are no doubt among the greatest safety tools that can save a person's life.

That's the good side of wireless technology because nothing has enhanced our sense of freedom like the cell phone, which, for many people, is increasingly becoming the hub of all their activities. The whole world seems to converge on this tiny device, which chirps at the most unexpected times and places. Millions and millions of people the world over use cell phones, walking and driving, without caring much for the hazards, such as car accidents due to distraction and, probably the worst of all, invasion of privacy due to interception. As the Internet becomes increasingly mobile and broadband, the networked cell phone, apart from becoming a security gadget, is also becoming a virtual self, a confidant for intimate moments—and a tool of interception and entrapment. Wireless communication is the least secure means of communication, yet it is indispensable to postmodern digital society. Some of the most intimate and distressed moments of conversation during September 11, 2001, took place on cell phones, and if the victims had camera phones, we would have seen it all live as we saw the hanging of the Iraqi dictator Saddam Hussein on 30 December 2006 on the Internet. When communication technology enhances our freedom and comfort, we forget the risks.

THE BARTNICKI CASE

Regardless of the risks, we cannot take technology back from the people once it adds to their comfort and enhances their sense of freedom. Cell phones, like cars, have become a permanent fixture of our society, and they are here to stay, because despite auto deaths, health hazards, and even invasion of privacy, they enable us to roam freely without losing touch with the rest of the world. To be free and connected at the same time is the greatest freedom. Of course, at some point privacy and personal freedom must be balanced with the public interest, as the U.S. Supreme Court decided in a case involving the interception of a cell phone conversation.[28]

During a 1993 contract negotiation in Pennsylvania, a teachers' union official negotiator, Gloria Bartnicki, had a cell phone conversation from her car with Anthony Kane, president of the local bargaining unit, condemning the school board's attitude toward their demand for more than a 3 percent raise. Kane said, "If they're not going to move, we're gonna have to go there, their homes . . . to blow off their front porches. We'll have to do work on some of those guys. . . . The rules are off." Bartnicki responded, "Exactly."[29] They did not realize that someone might be listening to their conversation.

An unknown person intercepted the cell conversation and handed the tape over anonymously to Jack Yocum, president of the taxpayers' association. Yocum was unsympathetic to the union's demand, and he passed the tape on to the school board and to a local radio station talk show host, Fred Vopper, who aired it several times during his broadcast. Then, the rest of the local media picked up the drumbeat and recycled the story. Although the news media were not involved in the interception of the cell phone conversation, the 1986 federal wiretap statute makes it illegal to disclose the content "of any wire, oral or electronic communication, knowing or having reason to know that the information was obtained through unlawful interception."[30] But when the story is a hot, sizzling scoop, the news media take a risk and publish it anyway, taking shelter under the people's right to know and the First Amendment. By the time the law catches up with them, the public does not recall what the case was about.

The union officials, Kane and Bartnicki, with the support of the Justice Department, sued the radio station and newspapers for civil damages for violating their privacy. The question as it got progressively framed in its legal journey to the Supreme Court was whether the First Amendment right of the media and the right of the people to know outweighs the rights of individuals to protected private speech, even though such speech might involve threats of violence.

Writing for the 6–3 majority, Justice John Paul Stevens said that the case implicates "conflicts between interests of the highest order," that is the interest in "full and free dissemination concerning public issues" and the interest in encouraging and protecting the freedom of private speech in the use of modern communications technology, namely, the cell phone. Since the conversation involved an issue of public concern, the teachers' contract negotiations, as well as a threat of violence, the Supreme Court absolved the media of any liability for using intercepted conversation on the ground that the First Amendment protects the press's freedom. The majority emphasized that it was a narrow ruling, based on the facts that the media did not participate in the interception of the conversation, that an illegally obtained conversation in itself is not illegal, though its use could be, and that the union officials were "voluntarily engaged in a public controversy" of great concern to the public in which threats were used and they had therefore opened themselves to greater public scrutiny than two persons in a private conversation.

The court distinguished a simple private speech from a "confidential conversation" between two people whose speech acts might have a great impact on the public. Would this ruling not eventually chill private speech on cell phones, which increasingly are being used by the American people? That was the point raised by Chief Justice William Rehnquist and the other two dissenters.[31] Even if the ruling of the Supreme Court was a narrow one, applicable to this specific case, it might be used as an excuse for tapping, for example, a private cell phone conversation between two companies' CEOs exploring merger possibilities, with the media reporting the intercepted conversation in the public interest. Privacy and confidentiality are not the same, though they fall into the same domain.

Bartnicki v. Vopper is not without a precedent. In the Pentagon Papers case, *New York Times Co. v. United States,* the Supreme Court in a 6–3 majority lifted the restraining orders against the *New York Times* and the *Washington Post* and ruled that they could publish secret documents about the Vietnam War, which they had received from a CIA employee who had copied them from classified files. Citing the Pentagon Papers case, Justice Stevens said that "the right of the press to publish information of great importance obtained from documents stolen by a third party" was of paramount importance and was supported by a landmark ruling in *New York Times v. Sullivan,* which established "our profound national commitment to the principle that debate on public issues should be uninhibited, robust and wide-open."[32] Illegal conduct does not diminish the importance of the First Amendment regarding a matter of public concern, in this case a dispute about teachers' salaries, which affected the lives of thousands of parents and children.

The cell phone makes it much easier to intercept conversation than a wired telephone because the interceptor does not have to bug a phone by either entering a premises or climbing a pole. Communications technology enhances freedom, but increased freedom creates more problems, and to solve them, you need more technology—or more social controls. That's how an endless spiral of technology creates endless problems for control. The solution to cell phone conversation protection seems to lie in strong encryption technology, but that might interfere with law enforcement's fight against terrorism and drug crimes, as we will explore in the following pages.

WIRELESS CULTURE

Broadcasting and communications technology played an important role in bringing down the tightly controlled and highly centralized Soviet Union. The Soviet Union's response to the North Atlantic Treaty Organization was the Warsaw Pact, but it had no solution for the information spread by nonviolent wireless and tape technology, which breached sovereign barriers. Walter Ulbricht, the former East German Communist leader of the 1960s, said in desperation, "The enemy of the people stands on the roof top."[33] There was no way of stopping people in Eastern Europe and the Soviet Union from listening to Western broadcasts, which played a significant role in gradually alienating them from Communist ideology.

Just as the Soviets could not prevent their citizens from listening to tales of ever-growing prosperity and rising living standards in the West, today no one can stop Americans from listening to each other's private conversations—thanks to the spread of cellular culture, which has made eavesdropping easy with a scanner modified for full-frequency use. Former FBI director J. Edgar Hoover wiretapped many important people's phones, including those of President John F. Kennedy and Attorney General Robert Kennedy, to exercise his power over them; that's how we know about some of their sexual escapades.[34] It was illegal, but Hoover thought he was above the law. Who would have dared to apprehend a supercop, one who knew too much? Today, surveillance devices are available to millions of people, who use them to listen to the police, fire departments, U.S. marshals, and drug-enforcement authorities, as well as to check on their spouses. It's not just the government and businesses that snoop on everyone; political rivals and even ordinary people intercept each other's cell phone conversations, as we saw in the Bartnicki case.

On December 21, 1999, a Florida resident, John Martin, and his wife, Alice, were Christmas shopping in Gainesville and aimlessly listening to their two-hundred-channel scanner when they happened to eavesdrop inadvertently on a conversation held by Representative John Boehner (R-OH), which was passing through their cellular scanning range. The Martins must have been intrigued when they heard former speaker of the house Newt Gingrich, Representative Boehner, and other GOP leaders discussing a strategy for dealing with Gingrich's ethics problem. Martin, a former local Democratic Party worker, taped the conversation and handed it over to Representative Jim McDermott (D-WA), a member of the House Ethics Committee. The taped conversation finally found its way to the *New York Times*, the newspaper of record claiming to publish "All the News That's Fit to Print," legal or illegal, to serve the public interest and maintain its posture of journalistic courage in being the foremost defender of the public's right to know.

The Martins beamed on national television as if they had performed a great heroic service, without knowing that their snooping would trigger an unintended consequence—an FBI probe into the conduct of the Democrats themselves, who were lunging for the ex-speaker's throat. Technology cuts many ways, but in a larger sense, the Martins' eavesdropping and taping of the Republican leaders' conversation might have served the national interest because the American people came to know the rest of the sordid story of ethical trespasses (tax evasion), the House reprimand (January 21, 1997), and Gingrich's subsequent decision not to return to Congress.

Electronic eavesdropping is against the law. The U.S. Supreme Court's ruling in *Katz v. United States*, as noted earlier, affirmed the people's right to be free from high-tech surveillance, even in a public telephone booth; nevertheless, the worst offenders have been the media and the government.[35] For instance, in the late 1960s, the State Board of Health of California and the Los Angeles district attorney entered into an arrangement with *Life* magazine to trap A. A. Dietemann, who was suspected of practicing medicine without a license. Jackie Metcalf, a reporter, and William Ray, a photographer, both with *Life* magazine, posed as husband and wife and went to see the "physician" for Metcalf's supposed ailment. While Dietemann examined the woman and talked about the prognosis, the conversation was secretly being relayed to a nearby van and taped. The recorded evidence led to Dietemann's conviction for illegal medical practice, which he did not challenge; but he sued *Life* for invasion of privacy and won damages. The judge, while announcing the damage award, said, "The First Amendment is not a license to trespass, to steal, or to intrude by electronic means into the precincts of another's home or office."[36] The media, nonetheless, have not stopped snooping for the sake of investigative reporting, which apologists justify as necessary for the public interest.

The protection of privacy from electronic eavesdropping was extended to include cell phone conversation in the 1986 Electronic Communication Privacy Act. Cell phone interception, though illegal, is not a felony and only carries a penalty of $500; however, disclosure of the intercepted conversation is a serious crime. If snooping on cell phones were a felony, there would be millions of cases every year.

Americans' love for openness arouses their suspicion when they find others being excessively secretive, which might explain why they want to demolish the walls of autocratic societies like Iran and North Korea; it makes them feel safe when the walls are down. Today, Americans are wirelessly unbound and are living in a transparent world not dissimilar to that which existed in times when people neither kept many secrets from each other, nor had the means to do so. But times have changed and unauthorized and unnecessary openness can lead to dangerous encounters and sometimes tragic consequences.

DANGEROUS ENCOUNTERS

Rebecca Schaeffer, a young actress who played a cuddly role in the television show *My Sister Sam*, became the victim of a fan-obsessed stalker who had obtained her address with the help of a private investigator using motor vehicle agency records.[37] Her tragic murder on July 18, 1989, eventually led Congress to enact the Drivers Privacy Protection Act in 1994, which prohibits motor vehicle agencies from giving out information about drivers without their consent. Some states protested that Congress exceeded its authority and lamented the abridgement of their "sovereignty," in short, the loss of income from selling drivers' private information in the marketplace.

The motor vehicle agency records about drivers contain vital and sensitive information, including their medical and disability information in some cases. The driver's license of a person is a window into his personal history and can be used to access a wealth of confidential information. If a criminal gains access to someone's record from the motor vehicle agency, that person can suffer a lot of harm. For the privilege of driving, which is a condition of survival, especially in rural states where public transportation is not available, states gather a lot more information than is necessary. A day might come when a person's organs may be taken out for transplant, if he forgot to check "I decline to make an anatomical gift," in the case of a fatal accident. The state might claim it owns the traffic victim's organs as it owns the personal information contained in the motor vehicle agency records, unless the person opts out, but most people do not pay attention to the opt-out option because nobody wants to think about dying in an accident and what should be done with their body parts.

In the age of superabundant information, when bits and bytes are rearranged, clustered, patterned, packaged, and merchandized for the marketplace, the sale of drivers' information has been a lucrative source of easy income for the states. Providers of mail-marketing services and owners of commercial databases pay for the names, addresses, telephone numbers, and medical and disability information contained in drivers' records. Criminals, too, have mined this rich source of information.

Deprived of this easy source of income, which runs into millions, some states challenged the law. In two cases, federal circuit courts upheld the law, while in another two, the law was struck down as unconstitutional. The U.S. Court of Appeals for the Fourth Circuit in Richmond, South Carolina, which voided the Drivers Privacy Protection Act of 1994 as unconstitutional, declared that the law violated "our system of dual sovereignty" by taking away the states' right to do whatever they want with information contained in drivers' records.[38] Such divisions in lower courts raise the serious question of whether Congress has exceeded its constitutional authority, a question that the U.S. Supreme Court must answer. The U.S. Justice Department, referring to Schaeffer's murder by the stalker, said in its Supreme Court appeal, "Evidence gathered by Congress revealed that that incident was similar to many other crimes in which stalkers, robbers and assailants had used motor vehicle records to locate, threaten and harm victims."[39]

The idea of dual sovereignty is a ridiculous fiction. Sovereignty implies not only jurisdiction over certain subjects and territories but also the right to raise an army and deal with foreign powers, which the states do not have. The Fourteenth Amendment leveled all states to a sub-servient status in the federal system. The amendment, which was passed after the Civil War in order to ensure that blacks enjoyed the same free-doms as whites, declares, "No state shall make or enforce any law which shall abridge the privileges or immunities of citizens of the United States." California, for example, cannot amend its constitution to limit the number of immigrants, but a sovereign state would be able to do so. On January 12, 2000, the U.S. Supreme Court affirmed the constitutional validity of the 1994 Driver's Privacy Protection Act pro-hibiting states from disclosing personal information contained in motor vehicle records.

In a 1999 case declaring California's two-tier welfare system un-constitutional because of its discrimination against newcomers, Jus-tice Stevens spoke eloquently: "Citizens of the United States, whether rich or poor, have the right to choose to be citizens 'of the State where they reside.' The states, however, do not have any right to select their citizens."[40]

Justice Stevens's remarks are not totally irrelevant to the question of the privacy of information contained in a driver's license. An American citizen owns the information, which is a form of property. No state has the right to deprive a citizen of his property without the due process of the Fourteenth Amendment. This is a fundamental right. At the same time, let's be clear that this is not only a privacy issue but a property is-sue as well, however trivial the amount involved in one's personal in-formation. In the United States, privacy, apart from the right to anonymity and to be left alone, is a form of property, and its appropria-tion without consent is a violation of property laws.

INTERNET ANONYMITY AND PRIVACY

While anonymous Internet messages can wreak havoc with the reputa-tion of individuals and corporations, as several recent court cases show, cyberanonymity, if used ethically and wisely, may have some virtues.[41] The Barnabus Christian Counseling Network of Ohio, which offers help for spiritual health, job stress, marital difficulties, aging issues, and emotional distress through online personal counseling, has been draw-ing a lot of attention.[42] Before a visitor begins to receive counseling, the service offers online testing in several areas, such as health personality, prejudice, ability to manage differences, burnout, and daily life stress. Based on what a person finds about him- or herself, the site offers guidance for managing personal problems and provides holistic healing methods, some of which are derived from the church's own teaching and others from commonsense relaxation practices, including some used by yoga teachers.

But if all this does not help, the Barnabus website offers two kinds of faith-based services, one through a counselor of one's choice in a private chat room and the other through a written online journal to which the counselor responds with assured privacy. The pastoral/faith-based counseling, though not free, is a significant cultural shift in the sense that whatever had been face to face in nature, from teaching to financial advice to discussion about one's relations with God, is now moving online. Eventually, all human activities will have their counterpart in the virtual world, and the interaction between the virtual and the physical will have a profound effect on our lives.

The Rev. Dr. Glenn Robitaille, the founder of the Barnabus Christian Counseling Network, started the online pastoral counseling to wean Christians from Internet pornography. As a pastoral counselor, he found that many decent members of his flock, who would hesitate to go to a video store to pick up an adult tape, were getting hooked on "sinful sexual activities" on the Net because it offered anonymity. The same Internet anonymity, however, might help a person cleanse his soul through confession and loud thinking before a trustworthy religious counselor. The Barnabus website has a strict privacy policy and swears by the confidentiality of its transactions; however, as every hacker knows, on the Internet every pixel has a tag, though sometimes we cannot see it. Reverend Robitaille might have been naively putting too much faith in the safety of the Internet when he said, "The level of honesty goes up with the level of anonymity."[43] The level of treachery also goes up with the level of anonymity. Anonymity is a many-faceted tool.

ANONYMOUS COMMUNICATION

If anonymity has the potential to lead us to honest, bold behavior, it does need some legal protection in the public interest. Most people are naturally timid and cannot tell the truth, even if they know it will help the public good. Some genuinely fear reprisals if they dare speak their minds. Whistleblowers, anonymous do-gooders, need to be protected. The anonymity of the Internet, if achieved properly through nontraceable methods such as those provided by anonymous remailers, can serve the public cause and expose corruption, especially in the case of popularly elected public officials, multinationals, and authoritarian regimes.

Anonymous communication on Internet bulletin boards can also help sufferers of sexually transmitted diseases, including AIDS, seek help without the fear of being socially ostracized or losing their jobs. It can help teenagers explore their sexual identities and learn to manage their libidos through an exchange of free, uninhibited discussion in friendly support groups. Nonprofit groups need to develop this unexplored region of cyberspace. But anonymity can also lead to social irresponsibility and even criminality by helping a person develop a dangerous persona. The case in point is the brilliant, but misguided, Theodore Kaczynski, the Unabomber, who developed his anonymous, murderous persona in the remoteness of Montana and sent mail bombs to people whom he considered the enemy of human civilization. Instead of sending mail bombs, he could have shared his paranoia about the excesses of modern-day high-tech civilization with others of his kind on the Internet under the cover of anonymity. Shared anonymity in a virtual community may be less dangerous than the isolated anonymity of an individual who feeds on his dangerous thoughts. For example, a New Hampshire politician, Tom Alciere, after getting elected to the state house of representatives, admitted that he had posted anonymous antipolice messages on the Net.[44] It is less dangerous hating cops online than killing them in reality. One day the Internet might develop into an anonymous forum for personal and collective catharsis.

For antisocial elements, anonymous Internet communication provides the best means of carrying out illegal and unethical activities, such as hate speech, electronic stalking, defamation, terrorism, violation of trade secrets and intellectual property rights, and, worst of all, child pornography.[45] From the point of view of the marketplace of ideas and free speech, anonymous communication deprives listeners of the opportunity of discussing the source of information and the motives of the communicator, without which, as Justice Hugo Black said, the recipients of the message might be "deceived by the belief that the information comes from a disinterested source."[46] Besides, anonymous Internet communication, if it is defamatory, discloses trade secrets, or violates copyright laws, can have worldwide repercussions and pose serious problems for the law-enforcement authorities who tackle such cases. If a pseudonymous Internet message posted on a Usenet group or community billboard wrongfully accuses a person of child molestation, there is no way the damage to that person's reputation can be repaired, and the person is deprived of his right to seek justice in an open trial under the Sixth Amendment, which entitles him to face his accuser. Without authentication of and accountability for public and private utterances, it is difficult to create trust, without which a free society or free market economy cannot be free for long.

But Internet anonymous communication has its virtues, especially in repressive regimes, where criticism of the government might lead to lifelong imprisonment or even death, and in corporations, where whistleblowers might lose their jobs. Internet communication potentially poses the greatest threat to autocratic regimes, though one wonders why so many are still in power. If anonymous Internet communication encourages people to seek help for their personal and intimate problems, including health issues such as cancer, sexually transmitted diseases, and AIDS, it could have a liberating influence on users and ultimately benefit society.

How serious is the threat of people using anonymous Internet communication to escape social responsibility for their utterances? It must be clearly understood that there is nothing like absolute anonymity because, in the long run, any message can be traced to its source, as the FBI did in the case of the Unabomber.[47] Internet anonymity is normally achieved through a chain of anonymous remailers, each of which replaces identifier information with a fake tag; if some remailers are based in foreign countries, tracing anonymous messages can become more difficult, though it is not impossible. A single weak link in the anonymous chain can break anonymity, as can a court order subpoenaing information.

The Church of Scientology, for example, issued a subpoena to identify the person who used remailers to distribute secret church teachings, which it claimed were copyrighted material.[48] Unless Internet remailers know that communication may violate copyright law or contain threats of violence or plans to commit acts of terrorism, they cannot be held liable for damages.

From the Federalist Papers' pseudonymous author, "Publius," to the Unabomber, Ted Kaczynski, all kinds of people have used anonymous speech to advance their causes. The courts in the United States, the ultimate guardian of free speech, have shown an ambiguous attitude toward anonymous speech. Kaczynski was convicted for his crimes, not for his anonymous writings published in the *New York Times*. In a 1958 ruling that had an important bearing on anonymous communication, the U.S. Supreme Court affirmed the right of the National Association for the Advancement of Colored People not to disclose its membership list to state authorities in order to protect the association from any retaliatory action on the part of the state.[49]

In a landmark case, *The New York Times v. Sullivan*, the U.S. Supreme Court unequivocally established the principle that public issues must be debated openly, without any inhibition, intimidation, or fear of punishment, even if the speakers make false utterances in the heat of debate. How far the principle of uninhibited political speech will be extended in the age of anonymous Internet communication is difficult to say, though in *McIntyre v. Ohio Elections Commission,* the U.S. Supreme Court ruled that "an author's decision to remain anonymous, like other decisions concerning omissions or additions to the content of publication, is an aspect of the freedom of speech protected by the First Amendment."[50] This was not a precedent-setting case in favor of anonymous speech because in another case, *First National Bank of Boston v. Bellotti,* the court ruled, "Identification of the source of advertising may be required as a means of disclosure, so that the people will be able to evaluate arguments to which they are being subjected."[51] It must be understood that the Supreme Court applies balancing tests among various fundamental rights (for example, between the First Amendment freedoms and the Sixth Amendment requirement for an impartial and fair trial for the accused). Besides, if the government shows a compelling reason that anonymous speech might endanger national security, the Court may allow anonymity to be broken. Considering various U.S. Supreme Court decisions, the free speech value of anonymous Internet communication in terms of the First Amendment depends upon the context of the communication. As A. Michael Froomkin says, "Despite its privileged position, political speech can be regulated given sufficient cause, especially if the regulation is content-neutral, as a regulation on anonymous speech would be."[52] Can medical privacy be regulated?

MEDICAL PRIVACY

The development of a unified nationwide database of patient health information is inevitable for several reasons, including health-care portability and medical research. There have been powerful market forces and lobbying interests at work, pressing Congress to legislate that they be given the right to know what transpires in the privacy of a clinic, claiming that their right to access medical data to strip-mine the American patient for information and build value-added databases for commercial purposes is more important than the patient's right to privacy. The market value of information increases, especially when placed in a relevant database, not only for direct marketers but also for companies that thrive on risk taking, for example, insurance companies. The marketplace predicated on the First Amendment, which is the bastion of core values, trumps privacy in the United States.

Every now and then, even the healthiest person falls sick. When over-the-counter remedies do not help and a visit to the family physician becomes necessary, it is presumed that no one will talk about the patient's ailment because nothing is supposed to leave the physician's examination room. But now patients face a new dilemma because once the information becomes accessible on the Internet, there is no guarantee that it will not be hacked into for fraud or blackmail. In 1999, the *Detroit News*, for example, reported that the personal and medical information of thousands of University of Michigan hospital patients, which included their Social Security numbers, employment information, birth dates, and diagnosis codes, was unintentionally released over the Internet.[53] With so much personal information available, anyone's credit history and identity could have been stolen and misused. Fortunately, in this case a student stumbled upon the unprotected database and alerted the authorities before any damage was done. Hospitals become protective of their patients' medical information and will decline requests for information even if the requester is asking about his own child. But they also naively believe that once the computer is turned off, the information is safe. Medical record confidentiality is a matter of serious concern to the American people, as several public polls show.[54]

Twila Brase, a registered nurse and president of Citizens for Choice in Health Care, made very insightful observations in the *Minnesota Physician*: "Most people make health care access decisions based on fear. Fear of death, fear of pain, fear of the unknown. If the fear of public exposure, political blackmail, employment discrimination or government surveillance is greater than the fear of potential disease and incapacity, the result could be incomplete patient disclosure, costly delay, skewed research results, and unwanted medical outcomes."[55] Many people hesitate to go to the doctor unless the ailment interferes with their work and they feel disabled, but once in the presence of the doctor, the doctor-patient mutual relationship rests on full disclosure and trust. Without receiving complete and correct health information from the patient, the doctor, apart from jeopardizing the patient's health and his own reputation, may also face the threat of medical malpractice, another reason for hospitals to gather more data about their patients.

COMPUTERIZED MEDICAL DATABASES

The doctor's dilemma about disclosure is that he is ethically bound by the solemn oath of Hippocrates (fourth century BC): "Whatsoever things I see or hear concerning the life of men, in my attendance on the sick or even apart therefrom, which ought not to be noised abroad, I will keep silence thereon, counting such things to be as sacred secrets." The new rules of the U.S. Department of Health and Human Services make it difficult for doctors to keep the "sacred secrets" of their patients because, effective April 14, 2003, the rules authorize doctors, hospitals, and other health-care providers to release medical information on patients for treatment or payment of claims without their written consent. All that the providers have to do under the new rules is notify patients of their privacy rights and make a "good faith effort" to get an acknowledgment from the patients that they have received the notice; written consent is not required.

In response to the public outcry against the dangers of computerized databases of private medical information, Congress authorized in 1996 the establishment of national privacy regulations under the Health Insurance Portability and Accountability Act, but left the details of the privacy rules to be worked out by the administration. While the new rules do away with the need for written consent by the patient before information is released for treatment and billing purposes, they prohibit marketing of the information for commercial purposes without the written consent of the person. Patients have the right to see and correct their medical records. They have the right to limit the release of information to an employer or law-enforcement authorities. For example, physicians or pharmacists cannot sell lists of patients taking prescription drugs for ailments to drug companies, who could use these lists to persuade users to switch brands.[56] These are the minimum standards, but state laws can give more protection to medical records.

There are legitimate uses of medical information, such as informing patients suffering from chronic ailments about new medical developments, evaluation of doctors for professional conduct, and detecting health-care frauds. The sharing of medical records, without patients' consent, among health-care providers and insurance companies may reduce paperwork and help in billing and identifying the most cost-effective treatment in some cases. If a heart attack patient had an emergency attack in another part of the country, the attending physician would have the patient's entire medical history and administer the necessary care without bothering about written consent. Large and consolidated databases would be very helpful in tracking certain diseases and how patients respond to specific drug treatments, which would help drug companies in conducting further research into their products.

PREVENTING MISUSE

An equally important question is how to prevent the misuse of medical information, for instance, an employer's firing a person based on medical information received from unauthorized sources or a person's being denied health insurance.[57] Private medical information can leak from computerized databases in many ways, including (1) accidental and unintentional disclosures by insiders, (2) browsing of patient information by employees through curiosity or malice, (3) sale of information by dishonest employees to database companies or to journalists looking for health information about politicians, (4) unauthorized access by outside intruders who sneak onto health-care premises and get into databases, and (5) disclosure by hackers and inside saboteurs, who do it for fun or malice.[58] When data get mixed up and doctors unknowingly make diagnoses and prescribe treatments based on compromised medical information, the consequences can be disastrous. The new rules do give patients the right to see their records and correct them, provided they know the correct information. But these are internal security problems, which can be minimized through checks and balances, firewalls, and the judicious use of encryption, especially for highly sensitive medical data, as not every piece of medical information is equally sensitive. Nonetheless, the privacy of the American patient has become negotiable under the rules of the marketplace.

Privacy has many functions. It gives people the opportunity to form a unique picture of themselves and ask that they be accepted as they present themselves. Based on this self-created persona, be it fiction or reality, they can do many things with their privacy. For example, they can market their image if they have achieved some notoriety in society as a sports personality, movie actor, or criminal. Excessive sunshine on a person's privacy, which digital profiling makes possible, dilutes a person's identity, hence, his or her marketability. Digital profilers can steal a person's persona. Marketable privacy is vulnerable to digital exposure, as is the stealth (call it privacy) that a person needs to commit crimes.

But privacy is also an emotional and psychological need, the need to be left alone so that a person can reformulate and re-create himself without anyone noticing the process. It is this creative aspect of privacy, the inner state of mind that allows a person to think about, tinker with, and re-create his personality, which is important not only to an individual but to society.

So the question is, How much privacy does a person need and for what purpose? In a democracy, such a question is inappropriate to ask because privacy is deemed not only a virtue in itself but also a fundamental right. In the long run, the question of privacy will be decided based on the core values of a society, be it Saudi Arabia, China, or the United States. A woman's desire to control her privacy by wearing a burqa or reveal herself by sporting a Victoria's Secret "Hearts on Fire Diamond Fantasy Bra" is a visual manifestation of the invisible power of the bastion of core values of the society she comes from.

Chapter Four

Surveillance in Cyberspace

The government's authority to invade a person's privacy at home and digital space for law-enforcement purposes is limited by the Supreme Court's interpretation of the Fourth Amendment to the U.S. Constitution:

> The right of the people to be secure in their persons, houses, papers, and effects, against unreasonable searches and seizures, shall not be violated, and no Warrants shall issue, but upon probable cause, supported by Oath or affirmation, and particularly describing the place to be searched, and the persons or things to be seized.

The passage of the first ten amendments to the U.S. Constitution in 1791 was, in part, a reaction against the excesses committed by European governments, particularly the British, especially against their own people. In time, however, the Bill of Rights in symbiosis with the free marketplace evolved into the controlling center of American society, the system's gravitational pull, its major cultural attraction, which began to define the character of the new nation, a new civilization. This was the first time that any society put constitutional limits on the authority of the government to compromise the people's fundamental rights. These rights were not granted by the government but were the people's natural rights endowed to them by some suprapower, the people themselves. The founding fathers deemed these rights to be fundamental and elemental to the human condition and American citizenship. A tripartite

government of divided powers, limiting one another's reach, accountable to one another and to the people, has proven a most efficient and
dynamic checks-and-balances, self-correcting, and self-renewing system that has guaranteed individual freedom and been the source of
American society's entrepreneurial spirit and vitality since its founding.

The burden of interpreting what the Constitution means at any given
time and under any given circumstance has fallen on the shoulders of
the U.S. Supreme Court, which the American people accept as the final
authority on everything that touches their lives. But the Court's interpretations are never carved in stone. They have changed according to
the changing mode of consciousness of the people and the impact of
new technology on the balance between liberty and security, between
the Bill of Rights and the government's constitutional obligation to
maintain and protect the system and its dynamics. Though in a given
case the U.S. Supreme Court majority gives the ruling as final, the
views of a dissenting judge today might become the court's majority
views in the future.[1] No judicial voice from the Supreme Court is ever
lost. The past creeps into the future and changes it.

SECURITY AND FREEDOM

National security and the maintenance of domestic order make up the
government's primary constitutional obligation, the fulfillment of
which sometimes conflicts with the fundamental rights of the people.
The government, limited in power by the Constitution, exercises its authority for search and seizure for law-enforcement purposes through
subpoenas, pen registers, tap-and-trace orders, wiretaps, and warrants.
A subpoena is a command to appear at a designated time and place to
give testimony that is deemed relevant to a legal proceeding and to hand
over documentary evidence, such as photographs, video and audio
tapes, and notes to authorities, provided the order meets the requirements of the Fourth Amendment. A prosecutor can also issue a subpoena directly or through a grand jury without judicial intervention,
though the information obtained may not be used as evidence. Pen register information is gathered by telephone companies for billing purposes and includes telephone numbers called from a particular phone,
as well as the date, time, and duration of the call. A tap-and-trace, on the

other hand, records incoming phone numbers to a particular telephone, which ordinarily a telephone company does not need, except when customers request caller ID. Pen register and tap-and-trace records obtained from telephone companies contain important information that can help law-enforcement authorities keep watch on a person under suspicion by looking at their communications traffic patterns. After independent evaluation, if a judge is satisfied that the government's request is relevant to an ongoing investigation, permission must be granted for access to suspects' pen register and tap-and-trace records.

The courts have held that the Fourth Amendment does not apply to pen register and tap-and-trace orders; therefore, the government need not show probable cause when making such a request, though tap-and-trace orders (caller ID) do have relevancy requirements under the law. Warrants, on the other hand, have stringent requirements under the Fourth Amendment; they cannot be issued without probable cause, and the request must specify the place of search and the materials to be seized. Law-enforcement authorities cannot go on fishing expeditions. Wiretaps, required for listening to ongoing conversations, must be done under court supervision. The probable cause showing in wiretap cases is more comprehensive than that for a warrant. While the Fourth Amendment is quite specific about requiring a search warrant before the police enter a home, in the case of telephony, which creates new space for people to do things, the U.S. Supreme Court had to take up the issue in order to give its interpretation as to what it means to be communicating on telephone.

The Bill of Rights, which empowers individuals to prevent the government from exceeding its constitutional authority, does not apply to private parties, such as business corporations, though some of these have power and influence exceeding the authority of a state government. As we become a database society, protecting individuals from the excesses of corporate power is going to be the biggest national challenge, which we will explore at length later in the discussion of corporate surveillance.

One Fourth Amendment protection sought by suspected criminals is the exclusion of illegally obtained evidence from their criminal trials so that they may enjoy their Sixth Amendment right to a fair trial. In 1791, when the Fourth Amendment was passed, a person's home was the most important source of information about him. Whatever information he

had to keep or hide was likely to be found there. Man's proverbial castle, the home was presumed to be the source of any information about,
or evidence of, a crime, and it was constitutionally protected from invasion by the government. Search and seizure of information including
papers and effects (triggering invasion of privacy) was house-and-loca-
tion based, and this mind-set of the Court, based on a literal reading of
the Fourth Amendment, continued through the later half of the twentieth century, even when the telephone and other forms of electronic communication were becoming widely prevalent. But in *Katz v. United
States* (1967), as discussed in chapter 3, the Court rejected the old idea
of privacy as limited to one's home and extended it to electronic space,
stating that the Fourth Amendment "protects people, not places."[2] The
Court held that law-enforcement authorities could not attach a listening
device to the outside of a phone booth without meeting the Fourth
Amendment requirements for probable cause and getting a search warrant, despite the fact that the telephone booth stood on public property.

REASONABLE EXPECTATION

The importance of the *Katz* case is that the Court developed a two-part
test in determining whether the individual has "a reasonable expectation
of privacy" as part of the Fourth Amendment's protection against government intrusion. As a first step, the individual must believe that his
communication is private, which is essentially a subjective assessment;
second, the society and its surrogate (the court) must accept that the expectation of privacy in the circumstance is reasonable. Thus, an individual's subjective expectation of privacy must be supported by societal norms and deemed reasonable. In actual practice, however, an
individual's subjective value of privacy reflects the prevalent societal
norms spread through popular culture. When something is portrayed re-
peatedly as normal in the media, people begin to accept it as natural, as
the way it ought to be. The social censor is silenced, and the individual
conscience stops being bothersome. This is how hegemony works in a
society.[3] People internalize what they see and read, day and night, in
movies, on television, and on the Internet, and in time, they unself-
consciously accept this as objective reality. Predominant ideas supported by the Supreme Court and reported in the media become the

standards, the natural environment in which individuals develop reasonable expectations. It makes life safer. This is how privacy and other individual rights are valued in the social mirror that reflects normalcy created by the guiding and controlling value system. So, even in one's home, for example, in the swimming pool in the backyard, a person's subjective expectation of privacy has no meaning if the society does not recognize it as legitimate. What a society recognizes as legitimate is described by the space created by its "bastion of attraction," the repository of its ultimate values, the inviolable core that holds the system in dynamic equilibrium. Privacy is not a stand-alone virtue. Privacy, freedom, and security form a continuum, an integrated circuit.

The idea of privacy differs from age to age and culture to culture. In the United States, the concept of property-based privacy rights at home was extended from real space to cyberspace, where individuals can reasonably expect privacy in their communication. Since a reasonable expectation of privacy is not a very precise standard that can be applied to every situation, especially when digital technology is adding new layers of communications to existing ones and creating new social spaces, much of the interpretation of privacy rights is left to the courts. For example, how much communication privacy protection under the Fourth Amendment should one expect in one's automobile? When a person invests in a mutual fund or makes a withdrawal from the bank, the information is not protected from the government's intrusion.[4] Nor can anyone claim privacy protection regarding the telephone numbers they dial from their home because they are disclosing the information to the phone company for personal or business purposes.[5] When a person shares personal information with a bank or telephone company, for example, any legitimate expectation of privacy in that information is lost, and the government, under certain circumstances, can access it, albeit not without crossing the Fourth Amendment barrier. A note of caution is necessary here because, while a person is not entitled to privacy regarding the telephone number he dials, he does not give up his expectation of privacy regarding the content of the message exchanged on the telephone company's wireless or wired-line system. The impact of the landmark *Katz* case has been so great that Congress acted to strike a new balance between an individual's renewed and extended right to privacy under the Fourth Amendment and the need of the government to meet its law-enforcement obligation.

In 1968 Congress enacted the Wiretap Act, which set parameters under which law-enforcement authorities could intercept the content of telephone communication for the purpose of criminal investigation.[6] The act requires that law-enforcement officials obtain a warrant from a judge by showing probable cause to believe that "an individual is committing, has committed, or is about to commit a particular offense" and that a wiretap is necessary because "normal investigative procedures have been tried and have failed or reasonably appear to be unlikely to succeed if tried or to be too dangerous."

The act does not prevent law enforcement from gathering telephone numbers a suspect might call, as mentioned earlier, and the act is limited to the content of communications being intercepted. The act does not apply if the parties do not have a reasonable expectation of privacy as sanctioned by prevailing norms.

Under the Foreign Intelligence Security Act (FISA), if the FBI or the CIA investigates or gathers intelligence about someone with ties to a foreign government, the act does not apply. In the investigation of a domestic crime, Wiretap Act standards do apply.

If one of the parties to a conversation has consented to surveillance, the act does not apply. While cell phones and display and voice pagers are protected under the act, transponders, beepers, and tone-only pagers are not. Although pen registers, which collect routine business information for telephone billing purposes, are not covered, taps-and-traces, which capture more than routine information (incoming calls), are covered under the act. This entire patchwork quilt of wiretapping laws intended to meet the needs of individual privacy and law enforcement had to be fixed as changes were brought about by further advances in communications technology.

The Electronic Communications Privacy Act (ECPA) of 1986 extended the wiretap privacy protections to include electronic communications, such as cell phones, computer-to-computer communication, and e-mail systems.[7] The punishment for unauthorized interception of private communications includes civil and criminal penalties. Law enforcement must obtain a search warrant for intercepting ongoing electronic communication, as well as stored wired and electronic communication for criminal-investigation purposes. The ECPA also provides a limited protection against pen register searches in the sense that law enforcement must show that the pen register information is relevant to the ongoing criminal investigation, as is the case with tap-and-trace-related searches.

To meet the challenge of emerging communications technologies, Congress amended both the Wiretap Act and the ECPA and enacted the Communications Assistance for Law Enforcement Act (CALEA) in 1994.[8] CALEA requires telephone companies and Internet Service Providers (ISPs) to update their digital equipment so that law enforcement can wiretap suspected communications for criminal-investigation purposes. A court order is required under CALEA for law enforcement to eavesdrop on or wiretap electronic or phone conversations on specific equipment used by a person under criminal investigation. The CALEA requirement is more stringent than the probable cause standard of the Fourth Amendment. In two cases, however, the courts have held that roving wiretaps allowing police to target suspects' communications in all locations did not violate the defendants' Fourth Amendment rights because surveillance was limited to identified speakers who were trying to thwart surveillance by changing locations.[9]

PUBLIC SOURCES

The Privacy Protection Act of 1980 was enacted to protect newsgathering media organizations and publishers from unauthorized search and seizure by law-enforcement authorities of the material on which reporters are working. The purpose was to allow reporters to work on investigative stories without interference by the government in the name of law and order. Under the Privacy Protection Act, law enforcement cannot search for and seize photographs, notes, or other materials without showing probable cause to the court that the materials sought are involved in the commission of a crime or in danger of being destroyed. The word "publisher" under the Privacy Protection Act includes not only traditional publishers of newspapers, books, broadcasts, or other forms of public communication, but also online systems that provide publishing services. The act extends its protection to the entire publishing system and is not limited to any specific part of the publisher's activity, such as printing or distribution, which has raised a serious question in the Internet age as to who is a publisher and who is not.

Keeping in mind the legal protections under various privacy laws, whenever any message is publicly accessible, it can be used by law-enforcement authorities for criminal-investigation purposes. For example, there is no reasonable expectation of privacy once I post my picture on YouTube or Flicker, and therefore my picture can be used as evidence in court if it meets the judicial standard of evidence. Chat rooms are deemed to be public forums, and the Fourth Amendment does not protect anything discussed there, even if it is a two-person chat in a private chat room. Pseudonymity does not give legal protection to users.[10] E-mail is treated like U.S. postal mail letters. A person who mails a sealed letter has a reasonable expectation of privacy, and the letter, in the process of transmission, is protected under the Fourth Amendment. However, once the recipient receives it, the sender loses control over the letter. The recipient can share its content with law enforcement voluntarily or may be obligated to do so under a court order after the Fourth Amendment requirements of establishing probable cause and getting a search warrant have been met.[11] Mass e-mails have no Fourth Amendment protection.

What if a device similar to the pen register or tap-and-trace, which normally records outgoing and incoming call numbers from a telephone, could be placed by law enforcement on an ISP's e-mail server to record header information containing the IP address of the sender and the recipient, as well as the subject line? Suppose the device enabled law enforcement to see the URL of every website a person had visited. Would permission to use this device, which sees more than numerical numbers, require merely meeting the "relevancy requirement" of a normal pen register order under the ECPA or would it necessitate getting a search warrant based on a probable cause? (While a pen register reveals only the telephone numbers a person has called, the URLs visited by him might show a pattern of his surfing behavior which could be used to profile him.) The U.S. Patriot Act makes such a question irrelevant in the name of national security.

Every time a new communications technology is developed, new space is created for the expression, storage, and distribution of ideas, enhancing individual freedom, which sometimes disturbs some people's privacy or threatens the government's ability to gather information for law-enforcement purposes.[12] The courts in the United States normally apply the balancing test, deciding in each case whether the rights of an individual outweigh the needs of law-enforcement authorities; in the process and through the judicial struggles, many new guidelines emerge.[13] Court rulings, precedents, and decisions constitute a kind of moving platform where the balance between individual rights and the government's constitutional obligations is established, until a new technology enters the marketplace, creates a new space for freedom, and disturbs the balance. New freedoms create the need for greater controls, sometimes for the better working of the marketplace and sometimes for national security. Threats to domestic and national security can create national paranoia, and frightened people might demand that the government take strong measures against the perpetrators. In an atmosphere of national crisis, civil liberties, unfortunately, seem less important, though in the long run their attrition might impact the free marketplace of goods and ideas.

THE PATRIOT ACT LOWERS BARRIERS

The immediate response to September 11 was the passage of an act that threatened to shrink the privacy rights of U.S. citizens in the digital age. The Uniting and Strengthening America by Providing Appropriate Tools Required to Intercept and Obstruct Terrorism Act (USA Patriot Act) was passed on October 26, 2001, and amended several statutes in order to enhance the power of law-enforcement authorities, including the FBI and the CIA, to do a better job of preventing terrorism.[14] The act enhances the government's authority to monitor individuals and financial transactions, to create a DNA data bank to identify terrorists, and to share grand jury information with law-enforcement authorities. It lowers the threshold for obtaining search warrants. Doing digital surveillance on the American people through wiretaps, search warrants, pen register and tap-and-trace orders, and subpoenas has become easy for law enforcement. The act also blurs the distinction between the intelligence-gathering and criminal-investigation branches of the Justice Department.

In 1978, Congress passed the Foreign Intelligence Surveillance Act (FISA), which authorizes the government to collect intelligence from foreign agents, foreign governments, and foreigners living in the United States. FISA's authority is far reaching; a search can be conducted without giving notice to the party under surveillance or without meeting any probable cause requirement. There is no discovery procedure under FISA; hence, the subject cannot get a copy of the FISA court order application detailing why the person is being put under surveillance. FISA has established a special court consisting of a panel of federal judges who conduct their proceedings in secret without any provision for an appeal process. The FISA court has no authority to monitor law enforcement once the court issues the order to intercept communication. Since FISA was enacted primarily and solely to gather foreign intelligence without adversely affecting the civil liberties of the American people, no serious debate took place. Now, however, the Patriot Act extends FISA's authority to criminal investigations also when foreign residents are involved, yet without meeting the stringent requirement of the showing of probable cause. FISA authority can be used to take away the civil liberties of foreigners residing in the United Sates without due process.

SPREADING TENTACLES

The Patriot Act enhances and extends government surveillance power over telecommunications technologies, including the Internet. Under the act, when law enforcement requests that the court issue a pen register/tap-and-trace order regarding a criminal investigation, the judge must issue it despite any reservation the bench may have. The act extends the pen register/tap-and-trace authority, previously limited to telephone communications, to "dialing, routing, and signaling" on the Internet because most of our communication activities now take place in cyberspace. Though the act prohibits gathering the content of communication with a pen register/tap-and-trace order, Internet routing information can reveal a lot about the websites a person visits. If a person sends an e-mail message to Al Jazeera, an Arab network, for example, law enforcement, using pen register/tap-and-trace authority under the Patriot Act, is supposed to look only at the e-mail address of the communication and not its content, though the two cannot be separated. The content of a telephone conversation can be separated from the number called; on the Internet, the information flows seamlessly. Since Web browsing for content and the routing information of a website's URL cannot be separated, law enforcement can easily profile a person based on routing information. Profiling a person without probable cause amounts to a violation of that individual's privacy rights because the Fourth Amendment protects him, regardless of where he happens to be, at home or in cyberspace. The Patriot Act has lowered the barriers for law enforcement to enter an individual's personal space. In cyberspace, the geographic location of an ISP has little meaning; therefore, a judge might issue a blank pen register/tap-and-trace order, which law enforcement can serve on a convenient ISP, in Colorado, New York, or Vermont, one less likely to challenge the order, since a message passes through several ISPs before it reaches the recipient.

Previously, the government could subpoena an ISP or website to get limited information such as a customer's name, address, mode of payment, and how long the service has been used. But now the act also allows authorities to subpoena customer information such as credit card and bank account numbers, which give more exact information about the user.[15] There is no check on how that information will be used. Under the Patriot Act, recorded telephone communication and stored e-mail and data can easily be obtained with a search warrant.[16] If someone sends an e-mail with an attached voice message, law enforcement needs only a search warrant to access it. Earlier, to get access to recorded voice mail, law enforcement needed a wiretap order, and accessing e-mail required a search warrant. According to the sunset provision, these amendments to the ECPA and the Wiretap Act were to end on December 31, 2005; that did not happen because the Patriot Act was renewed in 2006.[17]

USING SYSTEMS AS EYES AND EARS

The Patriot Act allows private online system owners, ISPs, and other service providers to monitor hackers and others trespassing on their systems, if they are authorized by the user of their system, provided such interception is related to the crime under investigation and limited to the hacker or trespasser.[18] A bank that suspects hacking into its computers cannot do the monitoring of its traffic, but it can ask its service provider to do so. The act gives some policing authorities to system owners, but it does not specify how far the system operator can go. This temporary

provision was supposed to sunset on December 31, 2005, but it was renewed.[19] The act authorizes ISPs to disclose voluntarily any information about their customers and their communication transactions, if they
believe that such information might relate to "an emergency involving
immediate risk of death or serious bodily injury to any person."[20] How
would an ISP suspect that one of its users might be dangerous unless
some sniffing and ethnic profiling had been done? An ISP might agree
to keep watch on some of its Arab-/Muslim-sounding customers or on
suspected child predators in chatline in order to perform monitoring services if required by law-enforcement authorities. Apart from adding
substantially to the cost of operations for ISPs, the act delegates law-enforcement functions to private parties, thus turning them into an arm of
law-enforcement, though without accountability.

Since cable companies provide Internet and telephony, the pen register/
tap-and-trace order provision of the Patriot Act applies to cable companies also, though the Cable Act of 1984 prohibits cable companies from
disclosing customers' personal information to law enforcement.[21] Based
on the Supreme Court's interpretation of the Fourth Amendment, law
enforcement must obtain a search warrant and give proper notice to the
person whose property is to be searched. In certain narrow circumstances, however, when there is a legitimate fear that giving such notice
may endanger someone's life or that the suspect might escape, the
timely notice requirement may be suspended. The Patriot Act makes it
easy for law enforcement to get a warrant without meeting the timely
notice requirement by showing that the notice would jeopardize the investigation.[22]

The earlier requirement under the Federal Rules of Criminal Procedure that law enforcement obtain a search warrant within the district
where a search is to occur has been broadened. To meet the terrorist
threat, the Patriot Act allows a search warrant to be issued for any jurisdiction anywhere in the nation where a terrorist activity might have
occurred or a suspected terrorist might have moved to. The provision is
very similar to nationwide search warrants for e-mail and stored data.
Similarly, the authority for nationwide search warrants for physical
searches of suspected terrorists' premises has been extended to their
telecommunications in the provision of a roving wiretap. Earlier, roving
wiretaps were issued under exceptional circumstances, for example,
when a suspect might not use the same phone again or might access the

Internet from different computers. That was for the investigation of criminal activities; the Patriot Act extends the roving wiretap authority to FISA for intelligence-gathering purposes without requiring that any probable cause be demonstrated to a secret FISA court.[23] Combining dialing, routing, and signaling information under pen register/tap-and-trace order authority with roving wiretap authority, the FBI can now use the excuse of apprehending terrorists to watch the digital society truly like Big Brother.

Surveillance may be performed not just on individuals and computers but on the whole system of communications. The concept of a reasonable expectation of privacy has been diluted, if not totally suspended, after the events of September 11, 2001. How far have we come from what Justice Louis Brandeis said in 1928 regarding telephony:

> The evil incident to invasion of the privacy of the telephone is far greater than that involved in tampering with the mails. Whenever a telephone line is tapped, the privacy of the persons at both ends of the line is invaded, and all conversation between them upon any subject, and although proper, confidential, and privileged, may be overheard. Moreover, the tapping of one man's telephone line involves the tapping of the telephone of every other person whom he may call, or who may call him. As a means of espionage, writs of assistance and general warrants are but puny instruments of tyranny and oppression when compared with wire tapping.[24]

DOMESTIC SPYING

As if the increased powers of information gathering under the Patriot Act were not enough, John Ashcroft, attorney general during the Bush administration from 2001 to 2005, announced on May 29, 2002, that restrictions on the FBI to conduct domestic spying in counterterrorism operations were being relaxed. Without any public debate or authorization from Congress, the government expanded the FBI's authority to monitor political groups, libraries, places of worship, and cyberspace. Restrictions on domestic spying were imposed on the FBI when it was disclosed that the bureau had been conducting domestic spying under Cointelpro (short for counterintelligence program) between 1956 and 1971. The program was targeted at violent groups, such as the Black Panthers, Nation of Islam, Ku Klux Klan, and American Nazi Party, and at nonviolent civil rights groups, such as the Southern Christian Leadership Conference led by the Rev. Martin Luther King Jr. To correct the criminal excesses of the Nixon administration, President Gerald Ford and his attorney general, Edward Levi, restricted FBI agents from conducting surveillance of individuals in public places, like libraries and town meetings, to gather information about individuals, unless such activities related to an ongoing criminal investigation. FBI agents were prohibited from going on fishing expeditions to gather intelligence about crimes that might be committed.

Now the FBI has been unchained and may perform preemptive surveillance on American citizens and residents. The new policy allows the FBI's regional offices to conduct counterterrorism investigations without prior sanction from headquarters in Washington, D.C. Without the obligation to show that a crime has been committed, FBI agents can now visit commercial websites, including online chat rooms, to gather information and develop leads about planned terrorist activities. The FBI is authorized to use the commercial data-mining and profiling services of marketing companies to conduct predictive intelligence gathering about events supposedly being planned by criminals and terrorists.

The new guidelines also authorize bureau agents to attend any public meeting or place of worship, such as mosques, for surveillance purposes. Since there is no expectation of privacy in public places, virtual or real, Fourth Amendment privacy concerns are not triggered, but an FBI agent's presence at a public debate might make the participants more cautious in their utterances, thus preventing the correction of wrong ideas in an open encounter. Places of worship are places of trust. If visitors fear that government spies have infiltrated these spiritual sanctuaries, their trust will be diminished.[25]

It is quite possible that the increased power of the government in the long run will lead to the diminishing of international terrorism in the United States, but its effect upon the personal freedom and privacy of an ordinary American citizen has not been thoroughly assessed.[26] By combining the powers of modern marketing tools for collecting data and profiling customers with the extended reach of surveillance by the federal government under the U.S. Patriot Act, there is the possibility of creating an environment of continuous and pervasive low-level vigilance in the United States, which the American people might have to get used to as unavoidable. And, in time, they might even accept surveillance as natural, unless it begins to diminish the innovative spirit without which the free marketplace cannot flourish.

The marketplace, however, is not concerned about surveillance, which is in fact an important tool for its efficient working. The marketplace thrives on surveillance and freedom, and balancing them in a dynamic system is quite a challenge.

CORPORATE SURVEILLANCE

Companies are engaged in a Darwinian struggle in the global marketplace. Surveillance is crucial for survival. When employees go to work for a competitor, they might be carrying important company knowledge with them. Even during times of economic slowdown, companies watch closely who is applying for jobs, especially in critical and strategic areas, and anyone found trying to cross over to a competitor may find cause to ponder the consequences of his actions.

It was a painful lesson learned by Richard Fraser, an independent Pennsylvania insurance sales representative with Nationwide Mutual Insurance, when he explored the possibilities of offering his services to the company's competitor by sending e-mail job queries. Fraser was not on the company's payroll, and in fact he had his own independent office; nevertheless, he was using the company's e-mail system, and his computer was networked with the company's server. In his lawsuit against the company filed after his services as an independent agent were severed, Fraser alleged that the company had violated his electronic privacy rights under federal laws, including the Wiretap Act and the Stored Communication Act.[27] The judge saw no merit in the case, however, because, even as an independent agent, he was expected to use the company's resources for company business. In the digital age, office workers must understand that a copy of all e-mail communication is retained on the company's server, the electronic equivalent of a filing cabinet that belongs to the company and which it has the right to scrutinize. E-mail is not the best method for employees to keep their online transactions confidential.

Companies not only have the right to snoop into e-mail, but they may also watch what employees do when they multitask on their desktops. In 1999, Blue Cross Blue Shield of Michigan fired several employees for violating the company's policy and written warning against e-mailing obscenities using the company's computers.[28] Employees' sending pornographic pictures and sexual jokes to colleagues via e-mail has become unacceptable workplace practice. Employers can ill afford to ignore the lewd online conduct of their workers; they do not want the company to be perceived as tolerating a hostile work environment, and, moreover, the legal cost of gross negligence can be enormous. The marketplace determines when surveillance is necessary to combat lewd online conduct, and companies are monitoring closely how their employees use their digital resources.

Several court decisions regarding workplace privacy have confirmed that, in the United States, employees have few privacy rights over their e-mail if it is stored in the company's system. Keeping in mind that the free marketplace stands at the core of American society, one cannot ignore companies' legitimate concerns, especially regarding the confidentiality of their trade secrets, knowledge and intellectual property base, and ongoing contractual negotiations. Pornographic and sexually harassing messages exchanged among employees might not only lead to expensive liabilities but, more importantly, damage a company's brand reputation. The concern is not new, but the speed with which transactions are conducted on the Internet has made businesses paranoid about what workers do in cyberspace. Apart from business, other most common activities that employees engage in online during office hours include visiting popular sports websites to check scores, online shopping, and stock trading. It is not unusual for workers to keep a chatline or instant messaging service open while doing business-related work. Multitasking in the workplace has always existed, but the Internet has created new temptations and dangers. Continuous restructuring and layoffs keep many working people constantly looking for new openings and networking. Since one never knows where the ax might fall, a culture of continuous chatter in cyberspace has entered the workplace.

Despite warnings against the misuse of company network resources, workers regard the office as more than a place of work because, offline and online, the worlds of work, home, and leisure are colliding and converging. The office has always been a place for socializing, romance, and friendship. And the home is no longer exclusively for the family. If workers are expected to carry their office home on their laptops or cell phones, why can they not carry their household chores on the laptop to their office? The question can hardly be ignored because the number of people who telecommute to work and have their home computers networked to their office server has been increasing steadily since the trend began three decades ago. Where does the right to privacy and freedom end for an employee when home and office computers are networked for company work?

The dilemma of the workplace does not end there because some employees need greater freedom and privacy than others. Idle chatter on the company network might be a necessary condition for creative behavior and may have to be tolerated in places that thrive on originality and creativity. The academic community, for example, uses university computers, but that should not give the universities the right to monitor their data streams. It would be counterproductive to take away academics' privacy rights and academic freedom, without which university campuses would become panoptical dungeons. The need for privacy and freedom is not the same for everyone.[29]

Even in the corporation, imposing the same digital rules on all employees might generate a coercive environment, which could have an adverse effect on productivity in the long run. Although the law may be on their side, it is not enough for employers to warn employees that their clickstreams and e-mails are being monitored because that is no way to build trust in a knowledge-based society. Since monitoring is essential for measuring and evaluating efficiency, preventing fraud, protecting intellectual property and trade secrets, and maintaining a conducive workplace environment, the policy should be clearly laid out but executed with flexibility. The courts favor employers because their thinking is property based; however, productivity lies in the hands of workers, which ironically gives corporations reason for monitoring and profiling their employees. According to an American Management Association survey, 78 percent of corporations monitor employees on the job, often without informing them.[30] Some of them hire outside investigating agencies to monitor their employees.[31]

Electronic surveillance by employers includes, in various degrees, monitoring Internet surfing, storing and reviewing e-mail messages and computer files, videoing employee job performance, and recording and reviewing telephone conversations and voice-mail messages. Corporate surveillance is not limited to employees; it also extends to customers through cyberprofiling.

CYBERPROFILING

Cyberspace is a repository of valuable information left behind by surfers that can be stored in databases for target-marketing and other valuable research purposes. Cyberprofiling is emerging as an important business tool for reaching the right customers through narrowly focused and targeted advertising and e-mail marketing. Since women do most of the domestic buying in the United States, data miners and cyberprofilers are concentrating on websites aimed at women.

Once cyberspace held so much promise for women, writes Professor Ann Bartow in the *University of San Francisco Law Review*, that it was the closest women could come to being "brains in boxes":

> In cyberspace, we would not be judged by our bodies. No one would know when we have bad hair days. We would not have to wear make-up and high heels. We could even be "men" without hormones or expensive surgery. Then we began shopping and chatting over the Internet. Shortly thereafter, we learned that anyone in cyberspace could ascertain our gender, ages, incomes, education levels, marital status, sizes, consumer purchase proclivities, aspects of our health, and employment histories, and the number, ages, and genders of our children, and that this information could be used to sell us goods and services. Now, instead of brains in boxes, we are "eyeballs with credit cards."[32]

The anonymity of cyberspace was supposed to enhance women's freedom and empower them vis-à-vis men, but instead of reaching a new threshold of freedom and equality, women are being robbed of their autonomy through surreptitious profiling. To a great extent, it is the marketplace, not morality, which controls the movement of the pendulum that determines spaces for freedom and privacy in the United States.

Only in the matter of high-expenditure items, like houses, cars, or vacations, a man might assert his power, though even in these domains the woman's voice is finally decisive. On the other hand, advertisers have known all along that they can control and manipulate women's desires for products from Victoria's Secret babydolls to the cereals they eat.

In the 1920s, radio began to develop as a mass medium with the potential to reach millions of women, most of whom did not have jobs and stayed at home with children and extended families. Companies like Procter & Gamble developed daytime serial dramas to entertain, entrance, and hold women in endless, intertwining, twisted storylines so that they would listen to soap commercials, giving rise to a unique narrative cultural form and effective mode of communicating with women all over the world. In the 1950s, when television started to dominate American homes, soap operas were transplanted into the new medium and have continued to retain their daytime popularity, despite the fact that, today, most women work because work is good and helps keep the household running. Today, soap operas dramatize women's problems, ranging from date rape, AIDS, and workplace harassment to raising children and keeping up with husbands in a culture where matrimony and divorce flip-flop like two sides of a coin. Those who succumb to the charm of slow-moving daytime dramas also find irresistible the temptations of Oprah Winfrey and her talk show, *O* magazine, and Oxygen Media, where TV and the Web converge into a shopping mall.[33] Women are extremely desirable in cyberspace because they control the purse and love to shop. The Home Shopping Network and other interactive television shopping malls are funded by the generosity of women.

There lies the future of cyberspace as a medium of e-commerce, which will require the building of high-quality websites where women feel comfortable and do not mind dropping valuable data that can be aggregated and collated into reliable individual profiles. Imagine a woman having her own personal boutique in cyberspace where the salesperson knows her tastes and preferences and where all her problems can be solved. So, when a woman enters iVillage.com or the Oxygen Media portal, she has the freedom to join a women's chat group and make new virtual friends; talk with an avatar; explore fitness, beauty, cuisine, telecommuting, and parenting; find advice about her job and tips about marriage, dating, and relationships; and publish her personal story on the Web.[34]

As women begin to feel comfortable in their personal cyberspaces, they will be scanned for all of their personal and intimate data, though unobtrusively. And what is true of women is equally true of other demographic groups. The free marketplace creates dynamic patterns of desire and the illusion of freedom, whether in cyberspace or real space. In another society, Saudi Arabia, a different dynamic system of core values creates altogether different personal behavior patterns, longings, and illusions of freedom and autonomy. When societies converge and collide and scan each other in cyberspace, who knows what can happen. There may develop a new dialectic instead of a clash of civilizations.

AUTOMATIC DATA SCANNING

Information passes through several computer systems before it reaches the person it is meant for. In the process of transmission, on the way to its destination, the information can be monitored, intercepted, and stored by any mediating system, without regard to the user's privacy or knowledge. The Internet is geographically unbounded.

Online tracking technologies, such as Web bugs, beacons, and other unique identifiers, enable a company to gather information for feedback purposes regarding the number of times a Web page has been visited, referring sites, entry and exit pages, and the geographical locations of visitors.[35] The technology records what kind of Internet browser a visitor uses and its configuration so that the aggregate information can be used to design the content of the website in a manner that attracts most visitors. It also helps track the movement of a visitor within a website and across other sites, enabling it to develop a pattern of visits over time, information that is used to profile the visitor. The tracing bug can track the effectiveness of a banner ad by matching online purchases a viewer makes after viewing it, for example. If a visitor uses a search engine to find information, the search strings from the search engine contain information about the user, which can be used for profiling purposes. The tracing bug can transfer previously collected demographic (e.g., gender, age, zip code) and personally identifiable information (e.g., name, address, phone number, e-mail address) about a user to a data-mining or marketing company for the purpose of profiling. The technology enables companies to share such data about visitors, which can be used for further value-added purposes. Profiling can reveal whether a person is black, Chinese, or Arab, for example, and whether he would buy a Ferrari or be satisfied with a Kia Rio compact sedan.

Some online activities are deemed open to public access, such as messages posted on newsgroups, bulletin boards, chat rooms, and mailing lists. There is no reasonable expectation of privacy in such online activities, and they can be accessed by anyone, even though such postings may contain one's e-mail address and other personal information. Opinions posted on newsgroups are archived in searchable databases and can be accessed years after they have been posted and forgotten by the user of the service. Even restricted online forums, for example, members-only bulletin boards and private chat rooms with passwords, are not absolutely private because a communicator can record and forward messages to anyone. Online service providers' member directories list all subscribers, and the information can be sold to direct marketers. The ECPA makes it illegal for anyone to intercept and disclose e-mail, though with certain exceptions. The service provider may view e-mail if it is suspected that the e-mailer might damage the system or harm another user of the service. The service provider may view and disclose the content if either party agrees to inspection or disclosure. If the e-mail system belongs to an employer, the employer has the right to examine the e-mail content of the company's employees, as was discussed earlier. The content of e-mail may not be disclosed to anyone unless one of the parties agrees, a court orders the disclosure, or the information is connected with the commissioning of a crime and is required by law-enforcement authorities. Under the rules, any service provider through whose system the e-mail passes en route to its destination can lawfully intercept e-mail messages if there is suspicion of an illegal activity.

Data-mining companies use browsing patterns and transaction information for target marketing. Technology enables online service providers to upgrade graphics and programs, and in the process, they can access information on a person's home computer without their knowledge. Once a computer is online, it becomes part of the system, and the system operator or a hacker can access its memory, where information privacy is preserved like a virtue, very vulnerable.

OBSESSIVE CONCERN

Polls show that people are concerned about digital privacy; nonetheless, the question is, Is digital privacy an obsession we should move beyond, or is it something more precious, a necessary human condition? Invasion of privacy in the digital age is nothing but unobtrusive data collection performed when a person surfs the Internet. Through data mining and profiling, the raw information can be turned into a valuable commodity by modern-day computing and marketing techniques, enabling the market to achieve greater efficiency. Profiling enables marketers to visualize consumers and create unique messages for them, with no face-to-face encounter. A most important feature of living in a digital society is that you can know a person intimately without meeting him or her personally, which raises a question: if data collection is crucial to e-commerce, how can the users be assured that surfing for business or pleasure will not erode their dignity and self-worth, which are normally associated with privacy? If the fear of losing digital privacy keeps people away from the Internet, e-commerce will not flourish, and the marketplace will not allow this state of affairs to continue. In the interest of the free marketplace, there may be no viable alternative but to protect privacy when a person visits a website, because there is a growing awareness that "the technology underlying the Internet is making it even easier and less expensive to gather, store, analyze, transmit and reuse personal information in ways that were unimaginable just a few years ago."[36]

Digital profiling, a function of artificial intelligence, uses databases to identify, segregate, and categorize information to make assumptions about individuals without anyone's ever having met them. The widely prevalent perception that databases exist in perpetuity without anyone's being able to correct them has the potential to create oppressive anxiety in some people, though most of us go on with our lives without ever worrying about being part of the data world unless trouble knocks at our door. In predigital days, profiling an individual from scattered databases required grueling hours of legwork and expense, but now it can be done with a keystroke. As profiling methodologies improve with digital technology, profiling will become instant and affordable even to individual users. Several companies specialize in data mining and profiling according to customers' needs, and this is one of the fastest-growing dotcom business areas. "Layering data and looking for the patterns within them is hard to beat as a means for connecting the dots," writes Stuart Brown in *Fortune*, whether the purpose is to target customers or apprehend terrorists.[37]

For the individual's part, sharing personal information may begin innocently and in good faith through online registrations, surveys, application forms, and transactions, as a necessary condition for going digital. What is the harm in disclosing e-mail addresses, postal addresses, and phone numbers? That's the thought that crosses the mind of a person in a hurry to surf a website. The Web creates a sense of urgency, as if we are trying to catch up with fleeting time. Some social networks, bulletin boards, and online chat rooms require personal disclosure for security reasons, but the visitor has no control over what happens to that initial information as it passes through digital processes. No one knows who the ultimate collector is, what the final shape of the information will be, or how it will be used, by whom, and for what purposes. Commercial data patterns might reveal a visitor to be a child molester, terrorist, psychopath, or entrepreneur.

It is easy to trust a local bank because we see the building and the people who are there every time we go. This sense of security can be misplaced, but it is there at least for the predigital generation. The digital generation can easily be persuaded to trust the one-to-one online relationship that an e-commerce website tries to foster. A Federal Trade Commission report noted that "with more detailed collection of data on a child, future e-mail solicitations may come from an animated character appearing on the child's screen, addressing him by name and urging him to purchase a specific product—perhaps a product over which a child lingered the last time he visited the site."[38] One of the biggest worries is that children, through the wily temptations of a digital avatar—remember Robert Browning's *The Pied Piper of Hamelin*—might be sucked into cyberspace, and parents might lose control over them as information, entertainment, and marketing come together seamlessly in cyberspace.

UNINTENDED DISCLOSURES

A fascinating aspect of the Internet from an e-commerce point of view is that even if a visitor refuses to disclose personal information, Web tracking programs, bugs, cookies, hidden navigators, and the like can ascertain the browser, computer, and the Internet address of visitors and keep a track of their clickstreams and mouse droppings as they move from one website to another. Based on that information, a profile of the user's preferences emerges for future marketing decisions. Future technologies will refine this process to make it e-commerce friendly. Unique identifiers embedded in computer chips and software can also keep track of the digital data involuntarily left in cyberspace by visitors. By combining information collected from clickstreams, mouse droppings, beacons, bugs, and cookies with offline databases, it is possible to extract extremely valuable intelligence for marketing and law-enforcement purposes.[39]

Websites' purpose in collecting data from their visitors is not intentionally malicious; they only want to know what aspects of their services and products appeal to visitors so that they can customize pages with appropriate advertisements and links when visitors come back to the site. There is, however, nothing to prevent them from selling the information to other data-mining companies, who might add or extract new value from the information for future target marketing, or even to law-enforcement if authorized under the new rules. Databases, fluid and dynamic and sometime flawed, are the pillars of postmodern digital civilization.

The general awareness about personal autonomy as a cherished value began at the end of nineteenth century, as mentioned earlier, with an article written by a couple of future U.S. Supreme Court justices, which was in fact a protest against the increasing tabloidization of American society made possible by yellow journalism based on a new camera technology and color printing and fueled by the market to maximize readership for advertisements. In time, through case law and legislation, the right to privacy developed four different facets, only one of which is directly related to the traditional concept of privacy about which the Boston brahmins, Samuel D. Warren and Louis D. Brandeis, fussed so much. The four torts of privacy are (1) appropriation of a person's name or likeness for commercial purposes, such as marketing and advertising, which is nothing but a property right when a celebrity becomes a brand name; (2) public disclosure of private facts directly related to one's professional reputation, job and livelihood, and marketability; (3) trespass, or intrusion into one's physical space, which has become an issue due to a heightened awareness of the self and is correlated to urbanization and the compression of space and time in the digital age; and (4) false-light exposure, which is when a person is portrayed in a embarrassing situation, unless he is paid to be so portrayed. The Internet as a medium of spontaneous and unfiltered communication is subject to all the four torts. Since cyberspace is a public sphere, whatever information a visitor gives up there can be used by anyone, unless the society puts a value on it.

DOSSIER SOCIETY

The question is, How far should we go to protect privacy and for what purpose? Can something be done to enhance e-commerce and at the same time protect individual privacy in cyberspace? As we discussed in chapter 3, there is no comprehensive privacy law in the United States that covers all aspects of cyberprivacy, including personal data, but a straitjacket law regarding digital privacy would choke the marketplace, and life would come to a grinding halt. The plethora of privacy laws intended to protect one or more of the aspects of personal information have been fashioned to meet contingencies as they arise.[40] As Paul Schwartz and Joel Reidenberg say, "The standards in place for fair information practices depend entirely on the context. In one context, such as direct marketing, personal information may have only limited protection against unfair use. In another context, such as employee record keeping, the same personal information may be subject to stringent legal and business controls."[41] Information privacy is a variable value; its variability is symbolic of the postmodern digital age.

George Orwell's nightmarish vision failed even in the Soviet Union and China. Why has the United States not become a dossier society? A dossier society in which details about every individual are kept in a central database would allow powerful officials in the government or corporations to control the behavior of people. The federal government has been collecting data on the private lives of the American people through the Internal Revenue Service, federal census, and housing, welfare, and other programs that require data collection and storage. Private corporations, too, have been collecting data through credit cards and other means. All that data collection apparently amounts to an assault on privacy, but there is no irrefutable evidence, in the form of repeated national surveys, ethnographic studies, or some other scientific method, proving that any harm has been done to the individual creativity, self-respect, and dignity of the American people.[42] Nevertheless, this should not diminish our legitimate concerns about corporate surveillance done for marketing purposes, including concerns about the following:

- Whether the data are current and accurate
- Whether they are subject to universal applications, that is, whether data collected for one purpose can be used to answer questions in other situations
- Whether the method of data collection inherently creates racial and ethnic bias and stereotyping
- Whether constant electronic surveillance via the Internet and other means tends to create excessively docile, self-observant, and self-critical behavior

- Whether such surveillance chills free speech and dissent through self-censorship

Listen to the legal scholar Karl D. Belgum's concerns:

> Preservation of free, unmonitored individual space may be essential to maintenance of a society based on the value of individualism. As a result, the loss of privacy may be harmful not only to sensitive individuals who feel aggrieved by its loss. It may also constitute harm to society itself, even if individual members of that society do not mind the loss, and willingly give up their privacy or trade it away for financial or other considerations. The logical consequence of this point of view is that society itself has an interest in preventing individuals from exposing themselves to too much loss of privacy.[43]

But how much loss of privacy is too much is difficult to say.

Harm can be done to society only if people, in addition to being constantly monitored, are also supervised and forced to correct themselves as was done in the Soviet Union or Mao's China. Neither society ever achieved total transparency despite all the dossiers. Speculatively, one might say that "total visibility infantilizes people" and "impoverishes their inner life and makes them vulnerable to oppression from without."[44] But there is no evidence that the United States is becoming a totally transparent society, or any likelihood that it will, because the same marketplace data-collection and profiling technology also holds sufficient incentives for entrepreneurs to develop technologies and websites that protect information privacy, provided there is a sufficient market for it. Some information privacy concerns are already being addressed by existing laws, though no law can ever create a foolproof privacy-protection system without choking society to death.

DATA OWNERSHIP

The idea that individuals should have the right to negotiate with websites to sell their personal information, such as their demographics, personal preferences, and whatever else the website is able to collect on its own, is not workable because information belongs to the organization that collects it per the copyright laws. Even if a person refuses to divulge his personal information, websites collect it automatically anyway. The Internet architecture ensures that surfing a website leads to some self-disclosure. Moreover, there is a First Amendment right to gather information from a public source. Assigning property rights to personal information left by a person visiting an e-commerce website would give brick-and-mortar businesses an unfair advantage because they do not operate under such restraints. However, it is possible for a limited exchange market to develop in personal information in the Internet environment; for example, companies might gather information by offering incentives to visitors to their sites, and those incentives will depend upon the amount of information disclosed. All information is not equally private. A person may not mind disclosing information about his leisure activities, for example, but might balk at disclosing age, income, and other personal choices. But once the information is disclosed, it is difficult to recall it or prevent the other party from processing it and adding value to it by matching and aggregating it with other similar information and marketing it to someone else. In the digital age, every piece of information is a building block for something else. It is impossible to predict the final shape of information or its interpretation.

A person's property rights in data cannot have the same market value as a celebrity's name, face, and other personal attributes, which can be used for the purpose of advertisement and product promotion. The value of online personal information lies not in to whom it belongs but in how it aggregates and patterns with other similar information. For example, a person's zip code information has little meaning unless it tells the information gatherer who else lives in that zip code area. The person has no bargaining power vis-à-vis a company interested in the information unless he has a unique piece of information about himself, and it has market value. Belgum thinks that the only potentially marketable personal information is that which tells about "the consumer's own subjective interest in being marketed a given product, service or subject matter"; since this information is fresh, it can be sold—if there is a buyer. What is the mechanism for selling this fresh personal information to the buyer? Should a person go to a website information broker and say, Look, I want to buy a set of golf clubs, and I was wondering how much I would get for disclosing personal information to the golf company. Packaging and commoditizing personal information for millions of Internet users, even with opt-in and opt-out licensing options, would create a legal nightmare.[45] The marketplace would not accept it.

What we need is a pragmatic approach—not to sell our privacy in the open market or to ban totally the collection of personal data lest we become a dossier society, a society under total surveillance, but to prevent the abuse of private information.[46] We need an approach that gives a measure of control over our personal information so that a balance is struck between the personal needs of the individual and e-commerce's development needs.

We cannot disregard the First Amendment's free speech guarantees, which include the freedom to gather information, even for e-commerce companies. Since all personal information is not of equal value, it does not need the same degree of protection. A person's street address or movie preferences do not have the same privacy value as their medical information. This calls for both self-regulation and some measure of oversight from the government as to what the reasonable expectations of the industry and of individuals ought to be in regard to data collection, the free flow of information, and how it should be used.[47] Building consumer confidence about Internet transactions will spur the growth of e-commerce and eventually benefit the consumer by creating more job opportunities and lower costs. Unnecessary fears about the coming of a 24/7 surveillance society in cyberspace will keep the consumer away from the global marketplace. Instead of giving individuals inalienable rights regarding personal information, websites, under government guidance, should create intelligent choices for users by educating them as to what it means to share some information. This should include the right to opt out and the consequences of doing so.

In order to build consumer confidence in online privacy and e-commerce, several reports have been published during the last decade, and some consensus has developed. Essentially, these reports maintain that to create a dynamic balance between online information privacy and the marketplace, it is important to adhere to the following recommendations:

- Personal data should not be collected without the knowledge of the user.
- The user should have the option to opt out of disclosing certain information, for example, sensitive medical information or information about young children.

- Information collected should be accurate, relevant to the purpose for which it is collected, and secure against any alteration or misappropriation.
- The user should have the right to review and correct the data.
- Websites should be held accountable for any improper use of the information and subject to an independent third-party audit for compliance with information policies.
- There should be a mechanism for dispute resolution.[48]

The emergence of guidelines for handling personal online information has raised some questions as to how to implement them. For example, is it adequate to put a privacy notice on a website, or should there be a blinker admonishing the visitor to read the privacy notice? How much consent or agreement is adequate, especially in regard to the collection of sensitive information regarding one's health or children or a third-party's use of the information? Should opt-in, which gives the highest protection, or opt-out, which is normally ignored, be the default rule in gathering information? The European Union Privacy Directive is much more protective of information privacy than U.S. policy guidelines. The European Directive requires each member state to enact privacy laws and appoint regulators to enforce them. According to the directive, individuals have an inalienable right to control their personal information: "Information providers should therefore not sell the personal data of their customers without their consent."[49] The EU directive also stipulates that data may not be transmitted to any country that does not meet European information-privacy standards. If the directive is enforced strictly, however, e-commerce between the United States and Europe will be adversely affected unless a worldwide information privacy standard is established.

In the postmodern digital age, information gathering can be disguised in any form and raises many privacy concerns. Most people are not sure how much privacy they need or what specific concerns they have. It seems, however, based on various surveys, that most privacy concerns are based on some vague fear that data might fall into wrong hands and the person might be exploited.

Apart from individuals' fear of identity theft, blackmail, and denial of access to mortgage and health insurance due to information abuse, there is also the anxiety that data mining can be used to pigeonhole people into specific categories for political and social purposes. The portrait of a person that emerges from the data mining of clickstreams and offline database sources might differ from his self-perception, and he may not like what he sees in the cybermirror. A person who regards himself as a conservative might find himself described as a bigot based on his surfing habits.

Surfing the Web freely is just like lying on a psychiatrist's couch and revealing oneself, but without the prospect of one's mental health being improved by the psychiatrist's healing powers.

OPT-IN AND OPT-OUT

If technology is capable of turning Web surfing into an involuntary, self-revelatory activity for the Web surfer, the question arises whether some measure of control should be given to the surfer. Of the two options we have today, opt-out and opt-in, from the market point of view, opt-out is more conducive to e-commerce because it encourages the free flow of information, the lifeblood of a free market economy. In most cases, opt-out is the default option; unless a person deliberately decides to opt out (by checking the box, which most people do not bother to do) of allowing information to be collected from him during surfing, the information-collecting websites work on the assumption that it is legitimate to use the information according to the declared privacy policy, however nebulous, posted on its site. Most of the users, however, do not read a website's policy, and even if they do, may not understand it fully. Opt-out gives the responsibility of saying no to the user (he has to check the box if he wants to disallow), who normally ignores it.

If users are given an opt-in option (i.e., a website may collect information during their surfing only if they check the box for opt-in), they may not do so out of fear. Most people's natural tendency is not to allow anyone to collect personal information unless it is absolutely essential. People hate to give away information to mortgage lenders, for example, but they have no choice. If companies were denied the information that the present opt-out default option gives them, they would be basing their marketing decision on inadequate information, which would raise costs and reduce efficiency. If the opt-in option were to become the norm, companies would have to incentivize visitors to their site to allow them to gather information, or perhaps they would have to exclude users who refused to reveal personal information from certain services. Opt-in is a very powerful tool for nondisclosure of personal information (because a person has to think before he checks the box to say yes as opposed to opt-out, where not checking the box automatically means yes) and might be used for some sensitive information, such as medical information. For cell phone users, for example, who do not want to receive direct marketing solicitations on their cell phones, opt-in may be the best choice. The most important issues for the digital society include deciding whether information collected for one purpose should be allowed to be combined with information collected for some other purpose, what further value can be extracted from the combination, and why it should not be done. No less important is the issue of whether corporations, for example, hospitals and insurance companies, should be prevented from sharing data with one another if data sharing leads to greater efficiencies without hurting the user. In the United States, the issue of information privacy is being decided by the principle of the free flow of information and the marketplace. That, however, raises the question of database security and the use of tools, such as encryption, and the collateral consequences.

ENCRYPTION AND SECURITY

Inventors and developers cannot control the use of technology once it is in the marketplace. Nor can anyone predict definitively who will benefit or be hurt as its use becomes widespread. The Internet, which has made global networked communication possible, began with open software design, free resource sharing, and unhindered public information. But as its potential for commerce and finance unfolded, Internet users' expectations changed. The Internet could not have remained entirely a public commons for long because the world of business needs not only openness but also security for transactions. It needs openness so that visitors to commercial websites will change from potential to active consumers, who then need security for finalizing transactions. The need for open access and security driven by the marketplace is shaping the Internet from e-commerce to YouTube.

Since sensitive personal communication, legal contracts, proprietary data, monetary transactions, and medical records are at risk on the

Internet and other digital media, online users have been demanding assurance of secure business transactions from the vendors of digital technologies. Private investigators and lawyers, for instance, can access a person's backup e-mail files and gather evidence for litigation. In response to the public cry for digital security, the industry began to look to encryption technology to scramble Internet communications. The military has been using strong encryption technology to shield its computers from hackers and other invasive network intruders, but that's nothing new because the military and the diplomatic corps have been the cradle of cryptology. To secure data, large, multinational, multilocale corporations use an intranet, an internal computer network that is segregated from the Internet by digital security firewalls designed to protect sensitive data and keep them within the organization. Large corporations can afford to have intranets, but for the millions of people who use the Internet, there is little security for their personal communications and business transactions. Without trusted communications, business cannot expand globally via the Internet.

Encryption technology, such as Pretty Good Privacy (PGP) developed by Phil Zimmermann on a double-key, public-private system, offers an acceptable level of secure personal communication and financial transactions for most Internet users. But that which aids Internet users in their quest for digital security and privacy diminishes the reach of law enforcers. Law-enforcement authorities demand that they be allowed to obtain lawful access to the plain text of encrypted communications and data because, without access and other safeguards, the lives of American people will be in danger. Although encryption technology holds the promise of giving Internet users freedom from the prying eyes of the government, it also gives terrorists, drug dealers, and criminals a free hand to cause death and destruction without fear of apprehension. In earlier times, electronic surveillance was much easier, but now, with the widespread availability of voice and data scramblers and encryption technology, law enforcement's technological edge has been blunted.

The case of Ramzi Yousef, who masterminded the bombing of the World Trade Center on February 26, 1993, has been cited in support of the need for access to encrypted communication. When Yousef was arrested in Pakistan on February 7, 1995, the FBI found on his laptop encrypted data files containing plans to blow up eleven U.S. airliners in Southeast Asia in a terrorist action; the FBI was able to decrypt the files and thus prevent the planned attack. A stronger encryption system would have thwarted the FBI in its operations against the terrorist. If strong encryption technology is built into every digital system, as the industry wants, cracking the code could become very difficult and take much longer, severely handicapping preemptive and preventive law-enforcement actions, especially in times of crisis. It might definitively shift the balance of power to criminals and terrorists, unless law enforcers are allowed to keep pace with decrypting capabilities.

Under the Clinton administration, FBI director Louis Freeh and Attorney General Janet Reno told the Federal Communications Commission that the telecommunications industry must take two essential steps to help law enforcers: build call-tracing capability into computer switchers so that participants in a conversation can be identified, and give law-enforcement authorities access to the private keys of encrypted communications. To prevent their misuse, Freeh proposed that the spare key might be placed with a third party, a bank for instance, to be used only under court order. The industry and civil-liberties groups vehemently opposed the Clipper chip/key escrow idea.

Once again, it is a question of balancing the government's constitutional obligations to guard the national security with the individual's right to digital privacy. The dilemma can be resolved by asking a simple question: how much digital security does a person need and for what purpose? A 56-bit encryption technology may suffice for ordinary credit card transactions, though financial institutions may require a much stronger encryption technology.[50] As computing power increases, however, according to Moore's law, encryption security will become weaker. Since crime, like commerce, is becoming global in operation, seeking international cooperation in regulating the sale of encryption technologies has become a crucial problem. Essentially, the struggle over encryption has moved away from the question of personal digital privacy to balancing the demands of the marketplace to do business in a secure environment and of the government to collect intelligence about criminals and terrorists in order to enhance security. Privacy advocates are never tired of telling us that a free society depends upon privacy to encourage artistic creativity, political discussion, and freedom of association—obviously an exaggerated claim because all of these activities can be conducted with limited privacy also.

ENCRYPTION AND PRIVACY

Opting out of digital civilization is not a choice unless one decides to live a life of total isolation. Communication and information transactions take place via the Internet and other digital devices, which bring all kinds of people, friends and strangers, together in cyberspace. There is no place to hide in cyberspace.

Traceability and interception of communication, along with speed and efficiency, are some of the most important characteristics of digital communication. Whether the ability to trace and intercept communication is a problem depends upon who the user is. A marketing company would like to trace the identities of visitors to its website but would not want any unauthorized party to intercept its own commercial transactions with its customers. Intelligence agencies and law-enforcement authorities need to intercept and trace the communications of people under their investigation for intelligence or evidence gathering for court procedures, so naturally they oppose the availability of strong encryption to the general public. Strong encryption is a relative term. Today's strong encryption will inevitably become weak in the future since computer power multiplies exponentially, making it easier to break codes.

Individual users want a private and secure communication medium, and since the Internet is becoming the medium of choice and necessity, the question is whether it can meet the challenge of rising expectations. If an Internet message is intercepted but cannot be read, the message remains both secure and private. To a great extent, encryption or cryptography can protect digital communications, be they in storage or in the process of transmission. The market-driven digital revolution has made encryption technology a necessary tool for the protection of commercial transactions as well as personal information, such as medical and financial data. But computing speed has an adverse effect on encryption strength.

Since cryptography is an unstable platform for security, the search for good encryption is a Sisyphean struggle. Essentially based on substitution of one message element with another, cryptography is "a transformation of a message that makes the message incomprehensible to anyone who is not in possession of secret information that is needed to restore the message to its normal plaintext or cleartext. The secret information is called the key, and its function is very similar to the function of a door key in a lock: it unlocks the message so that the recipient can read it."[51] The name "John Doe" could be encoded as Z-50-K-20-L-50-V according to a random allotment of numbers and letters to each alphabet letter, and no hacker would be able to decipher it unless he had the key or used brute computer power to break the code. The receiver and the sender use a single key, but the single-key system of encryption has its security drawbacks. Too many people might share the same key, and each time the key is passed from the key-management center to the user, the risk of code breaking increases. If the single key or the algorithm on which it is based is lost, the encrypted message will remain unread by the receiver, who then has to wait for the key to be replaced. In a crisis situation, such as national security, when decisions have to be taken in split seconds, delay could be fatal.

The encryption system that has been in wide use since 1970s was developed by Whitfield Diffie and Martin Hellman and uses two keys, a public key for encrypting and a corresponding private key for decrypting the message.[52] The public and private keys cannot be derived from each other. For example, if I want you to send me an encrypted letter using the dual-key system, I must give you my public key to encrypt the message before you send it to me. When I want to read the message, I use my private key to decrypt and open it. The transaction is then secure and private. Since I will not know whether the sender is a genuine person, I will insist upon other security measures, such as a digital signature, key certification from a certified authority against the danger of someone placing counterfeit keys in the Web directory of public keys, and digital fingerprints, especially if I receive an important document. To mitigate the paranoia of the digital age, we need to create a web of trust.

PGP is one of the most popular double-key systems used today; and since it "relies on the web of trust, key certifications, and fingerprints to help people assess the authenticity of public keys," one can be pretty sure of the security of commercial transactions and other important communications.[53] But security cannot be absolute. Some locks are easy to break, while others are difficult, but there is no lock that cannot be broken, and there is no key that cannot be duplicated. A computer key is a string of binary digits, and it is their combination, the code, that determines the strength of the key.

The strength of a computer key security system depends on how long the key is as well as on the power of computers. For some time, the U.S. government classified high-level encryption technology, 40-bit and more, as munitions, which could not be exported without a license. But in a society where the marketplace and free speech converge, the government could not keep a total ban on encryption technology. There have been contrary pressures with regard to dealing with encryption technology. For example, free speech and privacy advocates have argued that encrypted messages in principle have the protection of the First and the Fourth amendments, and the computer industry did not want to be shut out from the lucrative international market in encryption technology. Law-enforcement authorities, on the other hand, saw the unrestricted sale of encryption technology as hindering their efforts to deal with criminals and terrorists. The sale of high-powered, sophisticated encryption technologies, for example, the 128-bit key considered to be uncrackable at present, would threaten national security.[54] With the development of a supercomputing grid, such as TerraGrid, it should be possible to break the 128-bit key in a reasonable amount of time.[55] The argument does not end there, nevertheless.

GOVERNMENT'S INTEREST

The government's interest in encryption technology is quite substantial. National security and the maintenance of law and order are the government's constitutional obligations, which must be fulfilled so that the American people can enjoy the freedoms guaranteed under the Bill of Rights. National security is the most important official argument for having some backdoor access to networked digital communications. But overzealous law-enforcement officials on their own or under the direction of politically powerful people could exceed their authority and diminish individual freedoms. The courts have always applied the balancing test, permitting the government to encroach upon individual freedoms only when there is a compelling reason. If the government must invade a person's private communication, it must convince the court that there is a strong probability that a crime has been committed or is likely to be committed or that there is a serious threat to national security and therefore breaking into the person's physical premises or digital space will serve a limited purpose. But this argument is based on the assumption that breaking into digital or physical space is possible. If a person suspected of engaging in antinational or illegal activities builds and hides in an impenetrable underground structure that cannot be accessed by law-enforcement authorities, do Fourth Amendment privacy rights protect such an activity? If the 1993 siege of the Branch Davidians' Mount Carmel Center in Waco, Texas, by the FBI and the Bureau of Alcohol, Tobacco, and Firearms, which resulted in the death of eighty-two church members, is a precedent, one might conclude that no home or dwelling is absolutely protected by the Fourth Amendment. Nor should it be.

Building codes in the United States leave open the possibility of the police entering a house, if necessary. So far it has been possible for law-enforcement authorities, with court permission, to perform electronic surveillance on people under suspicion of committing violent crimes, terrorism, drug trafficking, and other illegal activities. But strong encryption technology has the potential to make it impossible for the government to tap quickly into communications between two or more people—for example, terrorists suspected of planning an attack on the national infrastructure—preventing law enforcement from preempting it.

In order to protect the lives and liberties of the people from external and internal threats, the government claims a legitimate stake in encryption technologies. But under the pressure of business and computer-industry lobbies, the government's policy posture changed from total control over encryption through the Clipper chip to a reluctant acceptance of encryption technology as a necessary tool for the enhancement of secure online trade and commerce.[56]

The potential global demand for strong encryption technology is huge, and American companies do not want to give up the growing encryption market to international competitors. Besides, for Internet commerce to increase in the United States and reach its full potential, unfettered, secure online communication, without the threat of government surveillance, is absolutely essential. Commercial transactions will be sped up, costs will come down, and other efficiencies will be achieved due to computer automation, provided corporate customers are assured that data are not being intercepted by anyone. Some have argued that strong encryption that excludes backdoor entry by the government via a spare key is a necessary condition for Internet commerce to flourish.

As the Internet becomes the dominant medium for global trade, the demand for high-end encryption technology products will increase, especially when buyers feel assured that the government has no access to their data and their communications are secure. One might also argue that strong encryption will increase mutual trust between strangers, without which business cannot be conducted. But the corporate scandals of 2002 involving Enron, WorldCom, Tyco, Adelphia, Global Crossing, and many others, show that much more is needed to create trust in business than encryption.

ENCRYPTION AND DIGITAL RIGHTS MANAGEMENT

Excessive surveillance, whatever form it takes, tends to engender behavior based upon fear and to create an obsession with correctness. In any case, the framers of the U.S. Constitution wanted to keep the government away from unwarranted searches of a person's private property, where most intimate and private communications used to take place. Electronic communication and the Internet, however, have extended the question of privacy to cyberspace, and the only way to protect privacy in a digital society, where communication mobility is on the rise, is through reasonably strong encryption that is universally available, provided that the government, in certain circumstances, is not totally excluded from it.

Yielding to the pressure from computer technology companies and civil libertarians, the Clinton administration on January 12, 2000, liberalized the encryption policy so that U.S. companies could compete to sell encryption products on the international market but without any provision for the government to have a backdoor access key. The marketplace won the battle of encryption. But the government's interests in controlling crime and taking preventive steps to thwart international terrorism cannot be ignored. Moreover, the privacy needs of individuals in relation to marketing companies have yet to be fully protected because encryption can do more than simply protect data and other forms of communications from access by unauthorized persons. Encryption can be a very efficient tool for digital rights management.

The decryption key can be designed in such a manner that, apart from opening an encrypted message, it can also determine how long the message can be read, whether the message can be forwarded to someone or downloaded, and whether the message will self-destruct after some time; all of this falls under the rubric of digital rights management, an issue of great interest since our culture is moving into digital channels and cyberspace. For example, a college e-textbook can be encrypted in a way that allows students to read it only for the fifteen weeks of the semester during which they lease the textbook. But the encryption code can also monitor and track the students' other surfing habits, for example, the music websites they visit. The e-textbook publisher not only makes money from the online textbooks but can also build a database to sell later to other marketers looking for profiles of young consumers of certain psychographics. The textbook website may require students to fill in a data form that includes the disclosure of personal information that could later be used for advertising purposes, another potential source of income for the publisher.

Considered in this way, encryption technology is not only a tool for securing communication for personal and commercial purposes but also a source of power for marketing companies to control the consumer. Depending upon the profile, a consumer might be given some information and deprived of other information. While the American Civil Liberties Union and Americans for Computer Privacy (a computer-industry lobby group) have come together in preventing the government from having a backdoor key for strong encryption technologies or from limiting their export for commercial purposes, these groups' interests eventually must diverge on the proper use of uncontrolled encryption by marketers in relation to data collection and the profiling of Web users.

Encryption technology has the potential to give businesses extraordinary control over visitors to their websites. The control comes through the encryption technology's power to lay a data siege on users without their knowledge. Just as conservationists trying to protect and regulate a herd of wild polar bears implant radio transmitters in the animals by numbing them first, commercial websites lure the users to their websites and then put digital tracers on them.

USES AND ABUSES

Internet communication, which has subsumed all forms of telecommunication, whether wired or wireless, mobile or fixed, analog or digital, audiovisual or written, encrypted or plaintext, is driving the postmodern digital age. Work and play, buying and selling, and social and cultural activities are increasingly being conducted in cyberspace. Cyberspace digital walls are porous and easily penetrable. Interception of communication, intelligence gathering, and spying have become easy and tempting for criminals, businesses, and, most of all, for government. Spying is an ancient activity, but before the telecommunications era, which began with the invention of the telegraph in 1837, spying was a long, arduous, and risky task. Now, a computer-savvy high school programmer or a college dropout can penetrate firewalls and tap into a digital communication network from his desktop; in the not-too-distant future, he will be able to do so from his mobile phone and let loose worms and bugs into the system, which can result in denial of service and other problems. Such an attack took place on May 4, 2000, in Manila, the Philippines, when twenty-four-year-old Onel de Guzman unleashed Love Bug, which crippled worldwide businesses, costing them billions of dollars.

When someone says, forget privacy because it is dead, he or she is spreading a dangerous half-truth because what technology takes away, it can also give us back. Cryptography, the ancient art and science of concealing information, developed in the twentieth century, especially during World War I, when radio was taken over by the U.S. government for military communication. The encryption of radio communication for military purposes became a compelling necessity during the war, but as with other military technologies, cryptography began to creep out from under military control to be developed for civilian purposes. The breakthrough that came with the double-key exchange system, which allows two parties to communicate securely without the intervention of a third party, also created a potential shift in the power relationship between the government, the individual, and the corporation.[57] Apart from protecting communication privacy for individual and business transactions, encryption can also help in establishing the true identities of the parties and whether their communication is authentic; thus, it can act as a weapon against disinformation.

The government's record in using its wiretapping power has been unsavory. Since the days of the protests over civil rights and Vietnam, law-enforcement authorities have wiretapped whomever they feared, including political opponents, journalists, editors, Supreme Court justices, and members of Congress, culminating in the Watergate break-ins that led to President Richard Nixon's resignation. The Church Committee appointed in 1975 to investigate the role of intelligence agencies in spying on citizens during and before the Nixon administration reported that the presidential administrations, including those of Harry Truman, Lyndon Johnson, John F. Kennedy, and Richard Nixon, had a long history of using national security as an excuse to wiretap their opponents without court authorization.[58] The Bush administration's record is no different, as the 2006 disclosure of the National Security Agency's practice of eavesdropping on Americans without court approval has revealed.

Despite the widespread use of wiretapping, both court-ordered and surreptitious, there is no evidence that wiretapping leads to a higher rate of criminal convictions. Nor is it indispensable to effective law enforcement. In many cases, as one judge put it, wiretapping becomes "mainly a crutch or a shorthand used by inefficient or lazy investigators."[59] Although wiretapping is expensive and time-consuming and its effect on crime prevention and conviction may be doubtful, in some cases, it may serve an important exploratory function in pursuing criminal and terrorist leads. Its function in intelligence gathering against terrorists cannot be minimized but needs to be regulated by the courts. However, cryptographic systems of "strategic capability, those capable of resisting concerted cryptanalysis by U.S. intelligence agencies," may pose a threat to national security. Many countries are developing technologies "for extracting information by active penetration" into the networked systems of friendly and hostile countries.[60]

When the government is given the power to invade privacy for law-enforcement and intelligence-gathering purposes, abuses can occur. It becomes difficult to hold the government accountable for privacy invasion unless a major break-in like Watergate occurs or a celebrity is involved.

One might say the best way to stop the government from abusing its power is to disable it from using that power. Cryptography can reinforce the Fourth Amendment privacy rights of the people as interpreted by the U.S. Supreme Court since the *Katz* case, in which the idea of the reasonable expectation of privacy was broadened to include wherever a person happens to be or whatever means of communication the person uses. The Fourth Amendment was erected as a wall against government's intrusiveness into people's lives, but technological advances have made that wall porous. Cryptography in the hands of an individual can make the wall impenetrable again. The question is whether it is desirable to make the wall so impenetrable that the government can never break it. For a healthy society to work, there has to be some level of interpenetration between the government and the people. Just as the government operates under the Sunshine laws and the Freedom of Information Act, individuals, too, with several exceptions, should live a life of some openness without giving up necessary autonomy. Neither the government nor individuals can be hermetically sealed from each other as society moves toward digital civilization. Restoring balance is important, however.

Existence in a digital society puts us into one or another interconnected database from which can emerge a detailed map of our diurnal deeds and misdeeds. Though there seems to be no escape from databases, public-key encryption technology does have the potential to restore some control to individuals and to protect their autonomy and privacy through secure communication. Secure communication is also the foundation for global e-commerce and freedom of expression, neither of which needs central management by the government or a third party of the encryption or escrow key. Of course, criminals will also use the tools of secure communication available for commerce and personal communication, but it is a question of the cost and benefit to society. Law-enforcement authorities have other means to apprehend criminal activities, more efficient than wiretapping communications, which may be encrypted.

Encrypted communication might have played some role in the September 11 terrorist attacks on the World Trade Center and the Pentagon. It is difficult to say how much al Qaeda's worldwide terrorist network, with its base in a primitive society like Afghanistan, depended upon encrypted communication. An escrow key would not have given law-enforcement authorities any predictive intelligence to prevent it. If terrorists want to use cryptography, they need not buy an American product. The United States has no control over cryptography technology available in the rest of the world, though this does not necessarily render it helpless because there are other means of achieving the same goal. On May 10, 1999, the FBI in cooperation with local police were able to bypass the new Jersey loan shark and gambling don Nicodemo Scarfo's encrypted files by using a keystroke logger that enabled them to get his pass-phrase.

PRIVACY AND TRUST

The search for privacy will be our major obsession in the digital age. The World Wide Consortium (W3C) has been developing the Platform for Privacy Practices (P3P), which might address some of the issues raised at present, but as data-collection technology becomes more unobtrusive, the level of concern will rise again. The stated purpose of P3P standards is to give the user a measure of control over how much information to disclose and when while doing online transactions; it also gives websites an opportunity to meet users' expectations. According to the W3C,

> At its most basic level, P3P is a standardized set of multiple-choice questions, covering all the major aspects of a Web site's privacy policies. Taken together, they present a clear snapshot of how a site handles personal information about its users. P3P-enabled Web sites make this information available in a standard, machine-readable format. P3P-enabled browsers can "read" this snapshot automatically and compare it to the consumer's own set of privacy preferences. P3P enhances user control by putting privacy policies where users can find them, in a form users can understand, and, most importantly, enables users to act on what they see.[61]

The W3C proposed standards do not offer general privacy protection to users because they are not enforceable. They simply make the privacy policy of a company transparent in nonlegalese, standard, machine-readable language so that users understand what they are giving up in order to use the site and do online business. It is not a negotiating instrument that a consumer can use to bargain with the site, but if it exceeds the law in certain areas, the site can be served with a notice of compliance.

Since online privacy and data-protection policies differ from one country to the next, the government of a country will encourage its consumers to perform online transactions with businesses that meet their privacy standards. If American businesses do not meet the privacy standards of European Union countries, for example, these businesses might suffer disadvantages vis-à-vis Japanese companies, which do meet the EU's stricter standards.[62] Eventually, the pressure of the global marketplace might help forge worldwide standards for data and privacy protection, but P3P cannot do this on its own. P3P also cannot catch businesses with deceptive data-collection practices. There is no way a user can be sure that a website is not collecting more information than its policy states. As a tool for assessing privacy standards and comparing them with the user's expectation of privacy or government-established standards of privacy, P3P can be helpful as a first step toward transparency.[63] If a company has no stated privacy policy, P3P can alert the consumer about this lack.

The TRUSTe program, an industry-supported privacy partnership to establish consumer confidence in online business transactions, also boasts that it was "designed expressly to ensure that your privacy is protected through open disclosure and to empower you to make informed choices."[64] The TRUSTe trademark is meant to assure that a company's website will inform visitors about the personal information that is being gathered; how the information will be used, with whom it will be shared, and the user's choices in this regard; the safeguards in place to protect information from loss and mutilation; and the provisions for updating and correcting inaccuracies in the information. TRUSTe also has the Children's Seal, which commits a website to obtaining verifiable parental consent to gather information about children under thirteen; to informing parents about what information will be collected, how it will be used, and the parents' right to participate in the activity in regard to the information collected; and to assuring that personally identifiable information collected from a child will not be distributed to third parties or posted on the site. Nor will a child under thirteen be lured with prizes or other awards to disclose more information than is necessary to participate in the activity. TRUSTe claims to "monitor licensees for compliance with program principles and posted privacy practices," including reviews by a certified public accounting firm. TRUSTe uses a verification system called "seeding," under which the organization submits personal information to verify whether the company is pursuing its stated privacy policy.[65]

Despite industry's fear of consumer reaction against privacy invasion, many people complain that the industry continues to collect personal information because "the day-to-day trickle of our data into the banks of direct marketers is the basis of the economy. . . . [T]he connection between our virtual selves and our potential as consumers is the economic basis of the current version of the Internet."[66] Collecting market information about consumer behavior from such varied sources as credit card transactions, phone orders, warranty cards, and application forms has been an indispensable part of doing business successfully. However, the Internet makes data collection automatic and more efficient for focused advertising and marketing. The speed of data collection and its myriad uses have created a diffused anxiety in the public mind as to how the data, especially personally identifiable information, might be used by commercial websites.

The free market economy depends on advertising; so does the marketplace of ideas. The *New York Times* and the *Wall Street Journal,* for example, deal with serious news and analysis, but they are supported by advertisements. Advertising does not determine which ideas presented in the media will survive in the future. To decry the advertisement model for the Internet is to question a core value of American society. In many cases, advertising is the most efficient vehicle for spreading ideas, some of which later turn out to be false and are eliminated. Changing the economic model of the Internet is not a viable strategy for combating invasive technologies, which indubitably monitor us without our consent in the digital environment in which most of our activities are increasingly taking place.

A combination of approaches that strike a balance between our need for privacy and companies' interest in data collection for improved marketing might work. A report by the Progress and Freedom Forum regarding the privacy practices and polices of commercial sites indicated the impact of consumers' serious privacy concerns when they go online. Indicating that the marketplace is not totally deaf to consumers' anxieties about their online privacy, the report concluded the following:

- Commercial websites were collecting less personally identifying information than they had been.
- The number of websites using third-party cookies that monitor consumer surfing across multiple sites had declined significantly.
- Most websites provided a complete and comprehensive statement about their privacy policies and practices.
- Most of the sites offered consumers the choice to share their personal information with third parties, though opt-in and opt-out consumer options were still very poor.
- The percentage of sites that offered fair information practices, such as notice, modified choice, and security, had increased, but P3P adoption by commercial websites was slow in catching up.[67]

It is rather surprising that despite fears about the abuse of personal information, only 3 percent of consumers read posted privacy policies, according to a Harris Interactive poll.[68] The reason is not hard to find. Online privacy notices more often confuse the reader with their lengthy legalese. More importantly, online users get impatient with such notices and want to go directly to the material they are looking for. In other words, they would rather have a legal and self-regulatory system protect their privacy than have to be vigilant every time they go online.

Surveillance is universal because it is an indispensable tool for the survival of a group, society, nation, or any self-aware system. Modes of surveillance, from the design of architectural spaces to tracking technologies, have been changing, but assessing what is happening, what might happen, and what preemptive steps might be taken to shape the future requires the development of a panoptic system that does not unnecessarily diminish civil liberties and retard individual creativity. The Patriot Act and dataveillance technologies have given enormous power to both the government and corporations to observe, assess, and track people and to profile them in order either to change their behavior or to prevent them from doing things that the authorities regard as inappropriate. Communications technologies, from the cell phone to encryption, have made the development of an intense panoptic culture inevitable. The challenge is to find a new balance between freedom and control that not only ensures survival but also fuels corporate growth and societal self-renewal, which ultimately depends upon individual autonomy, creativity, and self-expression. The marketplace of goods and ideas, which occupies the controlling center of American society, needs both surveillance and freedom.

Chapter Five

How the Marketplace Shapes Creativity

Reconciling the chaotic freedom of cyberspace, which makes sharing copyrighted files clickable, with the individual's right to benefit from his creation is a challenge. In one form or another, copyright has been protected since German printer Johannes Gutenberg reinvented and developed the movable-type printing press in 1436 to 1440, eventually making the mass publication of books possible.[1] Since the middle of the fifteenth century, tangible expression of ideas, artistic creations, and inventions have been gradually moving away from their free existence in the cultural commons to the marketplace, which has been expanding due to improved distribution systems and new communications technologies, such as the printing press and, later, photography, radio, and television. Copyright has been stimulated by the marketplace.

But the digital age has presented a serious challenge to copyright law, a social arrangement that has been protecting the rights of authors and other creative people and indirectly serving the expanding marketplace. Worldwide losses from intellectual property theft are estimated at more than $600 billion, of which the United States bears the greatest burden.[2] Other leading countries in the production of movies, music, and software have also been losing billions to piracy, but unlike the United States, countries with limited resources do not have the leverage of trade and diplomacy to persuade others to curb copyright thefts and piracy within their borders.[3] So, the question is whether the information superhighway will be a domain of pirates or the rights of creators will be reaffirmed in balance with the public interest within the boundaries of the marketplace. Those who want the society to return to the golden age of the public commons, including advocates of the Open Source movement, are challenging the controlling center of American society, the dynamic freedom of the marketplace that stimulates the expressive behavior and inventiveness of the people.

However, creators and intellectual property producers are not lone, helpless creatures. They have the backing of very powerful market forces, media giants who have been pouring millions of dollars into politicians' election funds to protect their interests. In September 1995, the Clinton administration issued a white paper, "Intellectual Property and the National Information Infrastructure," which admonished "that there was an increasing pressure from the public that information should be 'free and unencumbered,'" thus making the Copyright Act virtually ineffective and inapplicable to whatever happens in cyberspace.[4] In other words, users would like copyright infringement in cyberspace to be decriminalized so that they can upload, download, and even transform original, creative expression with impunity. If intellectual property laws were to be done away with, why would anyone create movies, write books, record music, or develop new software? Without the incentive of the marketplace, cyberspace would be choked with intellectual garbage masquerading as creativity. By protecting from infringement creators' expression of ideas in a tangible form, the Copyright Act benefits them and also enriches society.

It is important to remember that it is the form and mode of expression that are protected, not the data that give rise to the expression and artistic form. Like the Elizabethan playwrights, authors and creators today can freely borrow ideas and theories from anywhere, so long as they do not appropriate anyone's unique expression. Although it may at times seem difficult to separate form and content from each other, the marketplace and the judicial system have the capacity to discriminate creative expression from its factual content. The public interest, for instance, that of journalists, scholars, and libraries, is protected by "the fair use" provision of the Copyright Act, which is determined by the purpose and character of use. For example, if a music critic uses a song for a newspaper commentary, it is fair use, but if Coca-Cola were to turn that song into a commercial jingle, it would be an infringement. The fair use exception to copyright ownership implies that libraries may lend books, scholars and journalists may quote from others' works, and satirists may parody and poke fun at others' creations. It also allows people at home to use their videocassette recorders to record shows for later viewing (called time shifting), provided no commercial use is made of these recordings.

The white paper, which was introduced as a bill both in the House and the Senate in September 1995 and died with the 104th Congress (then was later reincarnated as the Sonny Bono Copyright Term Extension Act and was passed in 1998), warned that the benefits of the national information infrastructure "will not be realized if the education, information and entertainment products protected by intellectual property laws are not protected effectively." The Internet, which has been called a giant copying machine and makes anyone with a computer and a modem into a global publisher and distributor, has the potential to diminish intellectual property rights.[5]

On the surface, it might seem that the white paper simply reaffirmed the original intent of the Copyright Act when it stated, "Protection of works of authorship provides the stimulus for creativity, thus leading to the availability of works of literature, culture, art and entertainment that the public desires and that form the backbone of our economy and political discourse."[6] But the white paper also contained a new interpretation of the Copyright Act that regarded the downloading and making of a single copy of a Web file, even for browsing purposes, a potential copyright infringement, which raised the question as to whose interests would be best served. Can the marketplace reconcile these two diverse interests, that of authors/publishers to make money by exercising monopolistic control over their creations for a limited time (extended in 1998) and that of the public to encourage tinkering and creativity for self-renewal and growth?

FREEDOM TO CREATE AND THE PUBLIC INTEREST

What would have happened to Mickey Mouse if the Sonny Bono Copyright Term Extension Act had not extended the character's copyright? Mickey would have returned to humanity's cultural commons, where, for example, the stories of the Old Testament, the Arabian Nights, and Shakespeare reside, and perhaps thousands of digital-age creators would have been free to doodle and dawdle myriads of new Mickey creations. But that would have killed a brand and a marketing tool for a global media company, doing a grave disservice to the marketplace. The U.S. Supreme Court could not have ignored the marketplace unless the act had threatened national security or hindered the creativity and innovation on which the marketplace thrives. Justices of the Supreme Court unconsciously listen to the sound of freedom emerging from the controlling center of American society, which holds its abiding values.

It has been argued that Disney and other media conglomerates have established a stranglehold on intellectual property (through copyrights, trademarks and patents, and relentless lobbying) and would lose millions of dollars annually if the U.S. Supreme Court were to rule that the Sonny Bono Copyright Term Extension Act (1998) violated the Constitution's directive to Congress to promote the progress of the arts and science by giving authors monopoly over their works as an economic incentive, albeit for "limited times."[7] The Copyright Act of 1976 gave authors exclusive rights over their creations for the life of the author plus fifty years and to corporations for hired work for seventy-five years. The Sonny Bono Copyright Term Extension Act extends the duration of copyright protection retroactively for another twenty years, requiring Mickey to work and make money for Disney for two more decades. So far, there is no evidence that American creativity has begun to dry up because Mickey Mouse is an untouchable vassal of the Disney Kingdom.

This was the eleventh time over the last forty years that Congress had used its authority to extend the act. But this is the digital age and the value of time—a year, for instance—is much greater than it used to be, say, at the beginning of the last century, because due to the convergence and rapid pace of technological change, much more can be accomplished in the same amount of time. Time is a relative concept. Twenty years in the digital age is an eternity, so by extending copyright terms, Congress has given media companies and Hollywood copyrights almost in perpetuity instead of for "limited times" as specified by the Constitution. Intellectual property is not real estate that you can own forever. It has a different social function, that is, self-renewal and growth through creativity, but the U.S. Supreme Court was not convinced that copyright extension would diminish creativity, though rejecting the extension might have hurt the American marketplace. A Hollywood global brand would have disappeared overnight.

On the other hand, just imagine if a private trust had held the copyright to Shakespeare's works and insisted that his plays could be performed only under very strict licensing conditions, which included specific instructions as to how each scene should be enacted. That would have frozen the Bard. But since the plays belong to the cultural commons of mankind, there have been several versions of *Romeo and Juliet.* If Shakespeare has not been diminished through various interpretations, why not allow the same creative treatment to Mickey, Bugs Bunny, and others? That has been the argument against the act's extension, but the parallel is wrong because Mickey and Bugs are market brands. When a creative expression becomes a market brand, its status changes.

Walt Disney created Mickey Mouse from his previous creation, Steamboat Willie, in 1928, which in itself was based on the comedian Buster Keaton's Steamboat Bill.[8] That is how Walt Disney built his empire, according to Professor Lawrence Lessig of Stanford Law School, counsel of record for *Eldred v. Ashcroft,* the copyright case in which the Supreme Court affirmed Congress's constitutional right to amend the Constitution. Said Lessig, "Creativity and innovation always build on the past. . . . Walt [Disney] parroted feature-length mainstream films . . . taking works in the public domain, and not even in the public domain, and turning them into vastly greater, new creativity."[9] Of course, nothing can be created in a vacuum. New ideas emerge when a person plays upon, modifies, or rejects old ideas. One cannot create without messing with other people's ideas. The past should not be allowed to smother the future, otherwise society will become less free, as one can see is happening in some Middle Eastern countries where the written word of centuries past has frozen the minds of younger generations.

It is uncertain, however, whether the extended Copyright Act will eventually make the United States a less free, and consequently a less creative, society, though it must be acknowledged that without freedom and creativity, the United States, too, might one day become just like another doctrinaire society, beholden to one absolute vision. Copyright, which preserves the past by protecting tangible forms of expression, and the First Amendment, which creates the future by expanding freedom of speech and thought, do not exist in mutually exclusive domains. They share the same space and can extend each other's domains. Copyright and the First Amendment are parts of the system of freedom of expression and coexist in dynamic equilibrium. We cannot change one without adversely affecting the other; therefore, the extension of copyright term limits by twenty years, an eternity in the digital era, might put pressure on First Amendment freedoms. Without freedom to create, there will be nothing to copyright. Without freedom, digital civilization cannot survive.

COPYRIGHT AS A CENSORSHIP TOOL

The freedom of the Internet is threatening the established order, the powers that have survived on the command and control of ideas. On December 6, 1999, a Salt Lake City, Utah, federal judge, Tena Campbell, ordered a preliminary injunction barring Sandra and Jerald Tanner from posting the Internet addresses of a document pertaining to the Mormon Church on their website, Utah Lighthouse Ministry.[10] The Tanner website began with Psalms 43:3:

"O send out thy light and thy truth: let them lead me."

Born as Mormons, the Tanners gave up their religion and have ever since been persistent critics of their old church, also known as the Church of Jesus Christ of Latter-day Saints, which was started by a charismatic figure, Prophet Joseph Smith, in 1835 and now claims a worldwide following. The Tanners said, "The purpose of this site is to document problems with the claims of Mormonism and compare LDS doctrines with Christianity." In July 1999, they placed on their website seventeen pages of excerpts from a Mormon manual, the *Church Handbook of Instructions*, which they claimed to have received from an anonymous electronic source.

The Intellectual Reserve, a corporation that holds the copyright to the Mormon Church's intellectual property, complained in a lawsuit that by posting bootlegged excerpts on their website, the Tanners had infringed upon the church's copyright, and the church was awarded a temporary injunction against the site. Undeterred by legal threats, the Tanners subsequently published only hyperlinks on their website, which gave three Internet addresses where the Mormon handbook could be found. The church and Intellectual Reserve filed another complaint that the Tanners, by publishing the addresses of the websites that contained bootlegged copies, had engaged in contributory infringement. Contributory infringement amounts to aiding and abetting the person who has infringed on a copyright in the first place. The judge's ruling for temporary injunction barring the Tanners' Utah Lighthouse Ministry from publishing the Web addresses or links to sites not approved by the church hit on a unique feature of the Internet, the hyperlink.[11]

Without the freedom to link and surf freely in cyberspace, the Internet is not much of a medium because hypertext is at the heart of freedom. Including a link to a website or publishing a URL is no different from recommending that a reader go to a bookstore to find a title on a particular topic or an unauthorized celebrity biography. We do it with the hope that the bookstore sells legitimate books, not pirated copies. But even if the bookstore did carry a pirated title, would that make our recommendation to visit the bookstore illegal? There have been some interesting copyright infringement cases in the past, but they have only limited applications to the digital age. Consider *Zacchini v. Scripps-Howard Broadcasting Co.* (1977). The court ruled 5–4 that when, on its late evening newscast, a local television station showed in its entirety Hugo Zacchini's "human cannonball" act, in which he was shot from a cannon across a distance of two hundred feet into a net at a county fair in Ohio, the performance was not a fair use as a legitimate newscast; it was an unauthorized appropriation of the intellectual property.[12] In another case, a California company started doing business under the brand name "Here's Johnny Portable Toilet, Inc.," calling its products "the world's foremost commodian," in mock imitation of the famous standup comedian Johnny Carson, who sued for damages for misappropriation. The court ruled that Johnny Carson's identity was wrongfully capitalized upon because he had "a protected pecuniary interest in the commercial exploitation of his identity."[13] In *Allen v. Men's World Outlet, Inc.* (1988), the Federal District Court in New York ruled that the film director Woody Allen suffered unfair competition under the Lanham Act, when a magazine ad for a clothing company used a picture of Woody Allen's look-alike Phil Boroff with a clarinet in his hand. The ruling stated that this created confusion in the minds of the public as to whether Woody Allen, who also played the clarinet, was in fact associated with the clothing store.[14] These three cases were not so much about copyright issues as they were about unfair competition in the marketplace.

The Tanners' website, Utah Lighthouse Ministry, did not use excerpts from the Mormon Church handbook for any commercial exploitation. The Tanners are social critics, who used the church's own literature to question its fundamental assumptions. The use of the church material comes under the fair use clause of the Copyright Act and is further protected by the World Intellectual Property Organization, to which most countries, including the United States, are the signatories. Judge Campbell's preliminary injunction against unauthorized links and addresses was extremely threatening to the shared culture of the Internet. The use of copyright to silence critics is an increasing trend that threatens free speech. In a suit filed on February 8, 1995, the Church of Scientology tried to silence one of its critics, Dennis Erlich, a former Scientologist, claiming infringement of copyright for posting to an Internet newsgroup some of the works over which the church had a copyright. The church also accused the Internet service provider and a bulletin board service provider of contributory infringement.[15]

SWAPPING MUSIC AND EVERYTHING ELSE

The virtual and the real are interacting and converging to create a new synergy and new challenges and opportunities. But cyberspace cannot have its own set of laws and authority totally divorced from the laws of the real world, though it is equally fallacious to argue that cyberspace is nothing but an extension, or a mere appendage of, the physical world.

In every sphere of human activity, from popular culture to manufacturing, the Internet is destabilizing the old order, and until a new dynamic equilibrium emerges, there will be a lot of anxiety. In 1998, when an online music-sharing cybercompany, MP3.com, began distributing music to its customers using compression technology, which saved both downloading time and megabytes of disk space, music lovers were thrilled. They could not only listen to near-CD-quality music on the Internet, but they could also download it to their portable MP3 players. The company bought thousands of CDs and downloaded them onto its server in MP3 format, and all one had to do was open an account with the company, insert a CD into the drive as proof of ownership, and listen to music on the desktop or in portable form anytime, anywhere. But buying a CD was unnecessary because users could register a borrowed CD with MP3.com as their own, and the music would become part of their collections, even though they had not paid for the CD. Compression technology has made peer-to-peer music sharing on the Internet very convenient for the user, but convenience sometimes calls forth serious legal and ethical issues, which cannot be put on the back burner.

The music industry was up in arms against MP3.com because the file-sharing technology (MP3) threatened the existing business model that has been giving music companies complete control over their products. Since the industry could not fight the technology, the American Association of Recording Artists and major labels sued MP3.com in January 2000 for copyright violation. MP3.com argued that its new service was nothing but a music warehouse for its customers, or, you might say, a cooperative music library for contributing members. That was a specious argument because MP3.com did copy CDs into its server without permission, which violated copyright law.

At present, copyright laws governing the distribution of music in the United States are a nightmarish maze, involving recording companies, publishers, and artists' organizations, including the American Society of Composers, Authors, and Publishers (ASCAP), Broadcast Music, Inc. (BMI), and the Society of European Stage Authors and Composers (SESAC). Despite a plethora of laws, music piracy is a multi-billion-dollar international business, and MP3 technology made the matter worse. But there is no fighting a technology that makes life comfortable and enjoyable. MP3.com introduced two new services, which excited the whole music world but frightened the recording industry. Through its My.MP3.com personalization service, MP3.com introduced the Instant Listening Service, which allowed music lovers to purchase tracks from MP3.com partners, store them with MP3.com, and listen to the music whenever and wherever they wanted. For example, instead of carrying an entire collection of CDs to Singapore, a user could just access his account with MP3.com in a hotel room and listen to any music track, be it by Beethoven or the Beatles.

Another interesting service that MP3.com introduced was Beam-it. With this service, a person could insert a CD into a computer drive, and MP3.com would instantly register the tracks and place the CD file into a password-protected database. It was unnecessary for users to use their own disk space, but more than that, after they had registered CD tracks into their permanent account, they could pass on the CD to someone else, and the other person too could establish their own account with MP3.com. Would that have been infringing on anyone's intellectual property rights? Anyone can lend a book to a friend without infringing on the author's copyright, but they cannot do the same with a CD. The Recording Industry Association of America and Big Five record labels, Sony Music, Warner Music, EMI Recorded Music, Universal Music, and BMG Entertainment, sought "relief against an ongoing infringement of copyrights in the sound recordings on some 45,000 audio CDs."

When a person downloads his own stored music from the MP3.com server, is it the same as broadcasting or a public performance, which would necessitate licensing from the copyright holders? U.S. copyright law allows people to listen to music on any device they like, so they should be able to put a CD on an MP3.com server and listen to it whenever they want from wherever they happen to be. For most people, convenience is freedom and should be good for business, provided there is a good business model. Sony Music, Warner Music, EMI Recorded Music, and BMG Entertainment agreed to a settlement offered by MP3.com after the adverse April 2000 ruling against the company in a federal court, but Universal decided to pursue the case to the penalty phase and was awarded $250 million in damages.[16]

Napster, started in 1999 by a nineteen-year-old college dropout from Northeastern University in Boston, did not store music. Napster was much more revolutionary than MP3.com in its concept and presented a different kind of threat and challenge to the music industry and artists. It simply acted as a conduit between two people with music in MP3 format on their hard drives and let them download each other's songs without anyone's knowledge or permission. Napster was a music information portal with a central directory for a free peer-to-peer exchange of music. If anyone wanted to listen to a Beatles CD, Napster servers, in a manner of speaking, would find one for them, whether it was available in London or Hong Kong, so long as it was on someone's hard drive and was included in Napster's central directory. The recording industry and the heavy metal band Metallica sued Napster for copyright violation, which eventually put the company out of business; the same happened in the cases of other challengers, such as Kaaza, Grokster, and Morpheus, who took shelter either in the time-shifting argument of Sony Betamax or the user beware argument against copyright infringement.[17] Despite the fact that these companies used cutting-edge technology for music distribution, the court construed it as copyright infringement creating unfair competition in the marketplace.

Music file swapping nevertheless remains a global phenomenon because many people, especially those of the N-Generation, wonder why they should spend $16 on a CD when they can easily ask a friend to e-mail them an MP3 file attachment of the CD. The prevailing logic is that there is little difference between a friend in London's sending a person a book by next day delivery and another e-mailing a compressed music file attachment using an MP3 system. This is indeed a multi-million-dollar question for which the courts in the United States are struggling to find an answer. The availability of music from Apple's iTunes has created new, legitimate opportunities for listening to music. The marketplace will determine the future, compelling the music industry to come up with newer business models that will make illegal downloading less attractive. File swapping and sharing are at the heart of Internet culture, which the marketplace has yet to assimilate and monetize.

ENDLESS CHALLENGES TO
INTELLECTUAL PROPERTY RIGHTS

The recording industry had hardly finished coping with Napster's music-swapping system through a court challenge, when another one sprang up. A Norwegian teenager, Jon Johensen, developed a program to decrypt a movie recorded on a DVD.[18] The Content Scramble System (CSS) is used to encrypt movies on DVD, and only DVD players with the built-in key to decode them can play the movies. This seemed the perfect system to protect digital movies from pirates until the Norwegian teen came up with DeCSS to break the code so that a movie could be compressed and downloaded onto a PC hard drive and shared with anyone on the Internet. But the code-breaking software, DeCSS, violated the Digital Millennium Copyright Act enacted by the U.S. Congress in 1998 to expand the existing Copyright Act to protect expressive works, art, music, movies, books, and so forth, in digital formats in order to conform to the World Intellectual Property Organization Copyright Treaty. What seemed like another case of unauthorized decrypting software was transformed into a question of free speech under the First Amendment and of the Digital Millennium Copyright Act's constitutionality, attracting some of the top lawyers, scholars, and civil libertarians into a legal battle against Hollywood studios, which had the full backing of the recording industry and the Justice Department.

The trouble began when a hacker's online journal, *2600: The Hacker Quarterly*, posted DeCSS on its site and published links to other sites distributing the code-breaking software.[19] Perceiving a serious threat to their digital movie market, eight Hollywood studios challenged the website and its publisher, Eric Corley. Much as Napster had become a threat to the recording industry, DeCSS was perceived by Hollywood as a grave challenge to its existing business model.

The case raised a most important question. What is a computer code? Is it a control mechanism or a source of creativity and free speech? Or is it both? If it is a control mechanism, does it violate the fair use provision of the Copyright Act that allows critics, teachers, and students, for example, to use movie clips for education, parody, or criticism? If computer code is a form of free speech, then the code breaker's code, DeCSS, is also a form of constitutionally protected free speech, despite its potential threat to businesses dealing with movie distribution. Consider the case of Princeton University computer scientist Edward W. Felton and his colleagues, who in response to a challenge from the industry group's Secure Digital Music Initiative cracked its code for digital watermarks, then were threatened with legal action under the provisions of the Digital Millennium Copyright Act if they published their findings in a research journal.[20] How could an industry group ask university professors not to publish their critical research?[21]

The ruling of Judge Lewis A. Kaplan of a New York federal district court against the cyberjournal *2600*, barring it from posting copies of DeCSS and publishing links to sites that posted the code-breaking software, went before a three-judge appeals panel of the U.S. Court of Appeals, where Kathleen Sullivan, an authority on First Amendment issues, asked the court to throw out the case and declare the Digital Millennium Copyright Act unconstitutional.[22] The act prohibits the use of any technology or even trafficking in it in any form (Felton's code-breaking research paper, for example) that enables a user to bypass the encryption code protecting the copyrighted work; at the same time, it also prevents the fair use provision of the Copyright Act.

The fair use exception of the Copyright Act created a balance between the intellectual property rights of creators and the free speech rights of individuals to use material for educational and expressive purposes, for example, teaching, parody, or criticism. The Digital Millennium Copyright Act shifted the balance in favor of copyright holders. If the comedians of *Saturday Night Live* wanted to parody a movie by taking clips out of an encrypted DVD, they could not do it unless they broke the code with DeCSS. At present, a teacher, for example, can take excerpts from the videotape of a movie and use them in classroom discussions. Since encrypted movies on DVD make it impossible to make copies, encryption technology protects business interests but shrinks free speech. The code breaker DeCSS, on the other hand, enhances the possibilities of free speech but has the potential to damage business interests worth billions of dollars—no doubt a matter of great concern to the cultural marketplace.

The DeCSS case had another thorny issue. Can a website be prohibited from posting a hyperlink to another site, say, one hosted by someone in Norway that claims to enhance free speech but may also pose risks to someone's business interests or indirectly lead to another copyright-protected website, such as the Church of Scientology or Mormon websites? Technology is posing a challenge as to whether free expression and business interests can be balanced.

Dmitri Sklyarov, a Russian computer programmer, was arraigned on September 4, 2001, in a federal district court in California for violating the Digital Millennium Copyright Act, which, as stated earlier, makes it a crime to develop or sell technology to circumvent the digital encryption on copyrighted works, such as e-books, music, and movies, in the United States. Sklyarov worked for a Moscow-based software company, ElcomSoft, which developed Advanced e-Book Processor, a program that can decrypt Adobe Systems' electronic books. There is no Russian equivalent of the Digital Millennium Copyright Act; nor can American law be enforced in another country without its cooperation. So, the only thing that Adobe could do, apart from filing a complaint with the Justice Department, was to ask ElcomSoft to stop selling the program on the Internet. ElcomSoft had advised its online buyers to make payments to an American online payment company, Register Now, an act that made both companies accomplices in the criminal violation of the anticircumvention provision of the Digital Millennium Copyright Act.

In 2001, Sklyarov came to the United States to attend the hackers' convention in Las Vegas, Nevada, where he presented a paper, based on his graduate work, detailing how the Adobe e-book system meant to prevent piracy was defective and could be easily decrypted with a program developed by his company. The scholarly presentation at the hackers' convention was a protected form of speech under the First Amendment, but it also violated the Digital Millennium Copyright Act. The controversial act takes the traditional copyright law to its extreme; it not only aims at preventing piracy but also makes it a crime even to discuss an encryption program's vulnerabilities and how it can be broken. Under the law, Sklyarov and his company were charged with four counts of trafficking in illegal technology and one count of conspiracy to disseminate information for breaking the encryption system. What was Sklyarov's legal status? Was he a martyr for digital freedom, as his supporters describe him, or was he a foreign commercial hacker peddling illegal technology and seeking shelter under the First Amendment?

When too many people disobey a law, it becomes unenforceable, which obligates lawmakers to overhaul it. It is clear from recent cases that revamping has become necessary to upgrade the Digital Millennium Copyright Act, which gives too much power in its present form to intellectual property owners, mostly big corporations, against the public interest protected by the First Amendment.

Sklyarov was set free on $50,000 bail and was asked not to leave the country. On his arrest, there were widespread protests both offline and online, with T-shirts and bumper stickers turning Sklyarov into a martyr of free speech. Adobe, coming under tremendous pressure from several public interest groups, such as the Electronic Freedom Foundation, and many prominent academicians, withdrew its complaint. ElcomSoft had already stopped selling the decryption program in the United States, which made it convenient for Adobe to heed the public protests. The company's enthusiasm for the Digital Millennium Copyright Act remains undiminished, however. The Justice Department, on the other hand, not only refused to back off from the first case of Digital Millennium Copyright Act violation; it has established special units to fight computer hacking and intellectual property crimes.

The software industry, too, is coming up with various kinds of "trusted systems" to make cyberspace a safe place for corporate America to do business. Lawmakers and code makers are singing the same tune, both serving the marketplace, which cannot prosper unless it coexists, in a symbiotic relationship, with First Amendment freedoms.

Cyberspace is the new frontier for human freedom, with tremendous potential to open all closed societies. But it is also the gateway to the global marketplace and needs not only firewalls for protection but also regulated freedom. The Digital Millennium Copyright Act has upset the balance between copyright holders' right to make money from their intellectual and artistic creations and the people's right to make a fair use of the copyrighted material for legitimate purposes. It is time to redress the balance and make cyberspace a platform for freedom as well as a marketplace for honest and secure business. And that's quite a challenge.

A BRIEF HISTORY OF THE COPYRIGHT

Before Johannes Gutenberg developed the printing press, the mode of public communication, due to limited literacy, was primarily oral. In the preprinting era, the method of composing texts, such as stories, poems, legal documents, and rituals, was to speak aloud and give a kind of dictation to someone who could write, normally a person from the higher social classes or a monk in the case of church documents. Freedom of communication was limited to a few at the sites of power, be it ecclesiastical or royal, and was aimed at controlling social, political, and religious change in order to maintain a stable power structure. Printing and the spread of literacy hastened the pace of change by disseminating ideas rapidly to the unprivileged classes; people began to think on their own rather than acquiescing to top-down-controlled communication. Printing created the potential for mass communication in the form of books, pamphlets, and broadsheets based on freedom of thought rather than royal or ecclesiastical interpretations of truth, thereby threatening both the king and the clergy. To control the spread of seditious and blasphemous materials, Queen (Bloody) Mary I, daughter of King Henry VIII of England, granted in 1557 to the Stationers' Company power to exercise "perpetual lease" to censor publications, which continued until the royal decree expired in 1694 and the parliament refused to renew it. By that time, writing, printing, and publishing had created a new marketplace of ideas, goods, and services, involving high financial stakes and large business interests. The marketplace had begun to assert itself over "the divine right of kings" and the infallibility of the church. In 1710, when the Statute of Anne, the first copyright law, was enacted by the English parliament, it was not clear whose rights, those of authors or those of the publishers, were being protected. Censorship of seditious and blasphemous ideas had to be considered along with the expanding marketplace in the printed word. In 1774, when the House of

Lords interpreted the Statute of Anne, it ruled that the copyright belonged to authors, but for a limited period of fourteen years, renewable for another term of fourteen years, after which the work entered the public domain, where publishers could distribute it in any manner they liked or other writers could create new works from it. At the end of the eighteenth century, the notion of intellectual property had begun to take shape and was being strengthened through the instrument of the marketplace.

Although it seems as though the copyright clause of Article I of the U.S. Constitution ratified in 1788 is based on the Statute of Anne, the framers were focused more on the public interest than the rights of authors: "Congress shall have Power . . . [t]o promote the Progress of Science and useful Arts, . . . securing for limited Times to Authors and Inventors the exclusive Right to their respective Writings and Discoveries." It shows a tremendous shift in thought from political control to creativity and inventiveness. Instead of the government's controlling ideas to exercise power, Article I creates a property value for individuals' ideas and inventions so that society can benefit from them. In 1790, the U.S. Congress enacted the first copyright statute that granted copyright protection for fourteen years renewable for another term of fourteen years. The statute was amended in 1834 to include music as a protected category and to increase the initial copyright protection period to twenty-eight years, though the renewal term remained unchanged at fourteen years. The same year, the U.S. Supreme Court made a landmark ruling in *Wheaton v. Peters*, affirming Congress's constitutional right to grant authors exclusive rights to their creations for a limited period. The Court at that time did not consider the meaning of "limited times" or the nexus between copyright monopolies and the First Amendment, which has become a serious issue today.

The copyright statute underwent substantial revisions in 1870 and 1909, primarily to expand the list of protected forms of creative expression to include plays, photographs, engravings, paintings, and drawings, as they entered the marketplace and held the promise of making money for their creators. New communications technologies, however, including motion pictures, radio, and television, were expanding speech and creating newer forms of mass communication, which the copyright statute was powerless to protect. Congress made another major revision of the statute and finally adopted the Copyright Act of 1976, which not only expanded the rights of authors in the age of mass media but also balanced those rights with the public interest through the doctrine of fair use. The act expanded copyright duration to the life of the author plus fifty years and to seventy-five years for work for hire. The act protects "original works of authorship, fixed in any tangible medium of expression," including literary works; musical works and their accompanying words; dramatic works and their accompanying music; pantomimes and choreography; pictorial, graphic, and sculptural works; movies and audiovisual works; and sound recordings. Computer programs are considered literary works and are thus protected under the copyright statute, but users are allowed to make a backup copy of a program in case the original program is damaged.

The statute gives authors and creators exclusive control over their work and includes the right to reproduce their work; the preparation of derivatives based on their work; the distribution, sale, transfer, lease, or rental of their work; performance rights; and the public display of their work. But the right is limited to the manner of tangible expression of the work, not the raw ideas or the information on which the work is based. Titles, historical facts, names, short phrases, slogans, familiar symbols or design, procedures, methods, systems, processes, concepts, discoveries, and devices are not copyrightable, though their descriptions, explanations, and illustrations, which are expressive in nature, are copyrightable. To be copyrightable, the work should be original but not necessarily unique or novel.[23]

HOW LONG IS A "LIMITED TIME"?

With each periodic revision of the copyright law, Congress extended the duration of the copyright without any constitutional challenge until the most recent extension, the Sonny Bono Copyright Term Extension Act of 1998, which added another twenty years to the 1976 Copyright Act. Now, the act protects copyright for the life of the author plus seventy years and for ninety-five years for hired work. Since most of the creative work done nowadays is for the consumption of media companies that serve the global market, the extension of copyright is meant to benefit media conglomerates rather than to serve the public interest. This shift has been taking place gradually. Consider, for example, Justice Potter Stewart's opinion in *Twentieth Century Music Corporation v. Aiken* (1975): "The limited scope of the copyright holder's statutory monopoly, like the limited copyright duration required by the Constitution, reflects a balance of competing claims upon the public interest. . . . The immediate effect of our copyright law is to secure a fair return for an 'author's' creative labor. But the ultimate aim is, by this incentive, to stimulate artistic creativity for the general public good."[24]

The Sonny Bono Copyright Term Extension Act, which extends publishers' rights to ninety-five years, ignores the general public good. There is no empirical evidence that each copyright extension has increased creativity or benefited the author, although it has certainly profited media companies. Had the term not been extended, works like *Rhapsody in Blue* and characters like Mickey Mouse, for example, might have gone into the public domain, though it is also difficult to say whether that would have led to greater creativity.

FAIR USE

An interesting aspect of the Copyright Act of 1976 is that it balances an author's financial interest in his or her work with the public interest through the doctrine of fair use. Section 107 of the statute states that "the fair use of a copyrighted work . . . for purposes such as criticism, comment, news, reporting, teaching (including multiple copies for classroom use), scholarship or research, is not an infringement of copyright." A determination of whether the use of copyrighted material falls under the provision of fair use, however, must consider in totality several factors, including whether the purpose of the use is commercial or noncommercial-educational; the nature of the copyrighted work, for example, whether the work is published or unpublished (unpublished works have greater protection); the amount of work used as a portion of the total work (a few phrases from a song may be an unfair use, while two hundred words from a two-hundred-page book may be a fair use); and its measurable effect upon the market of the copyrighted work, that is, the financial loss that the copyright holder might suffer.[25]

Congress could not have foreseen how the fair use clause of the Copyright Act of 1976 would accommodate the demands of new, user-friendly, inexpensive video communications technologies, like Sony's Betamax in 1979, which enabled viewers to record television programs for later viewing. With Betamax, viewers could turn their homes into recording studios and build their own libraries of classic movies, sports events, and favorite television programs, which Sony hyped in its advertisement campaign to create a market for the new freedom technology. Threatened by the new technology and fearing the loss of millions of dollars in revenue due to piracy, Universal City Studio and Walt Disney Productions went to court to seek an injunction to stop further sales of Betamax on the grounds that the video recorder was a tool for copyright infringement. When the case finally reached the Supreme Court, a 5–4 majority ruled that copying programs for one's personal use is not an infringement but a fair use and added a new expression, "time-shifting," to the legal parlance. Writing for the majority, Justice John Paul Stevens said, "To the extent that time-shifting expands public access to freely broadcast television programs, it yields societal benefits. . . . Concededly that interest is not unlimited. But it supports an interpretation of the concept of 'fair use' that requires the copy holder to demonstrate some likelihood of harm before he may condemn a private act of time-shifting as a violation of federal law."[26] The burden of proof that harm would be done by the new technology was left to Hollywood.

The Supreme Court's landmark decision in *Universal City* has had a tremendous impact on all areas of communications, such as computer programs, e-mail, movies, and audio recordings, where technology makes it easy to copy and blend sound tracks, songs, texts, films, photographs, and graphics to produce new media products. The Court could not have stopped the emergence of the copying technology, but by expanding the boundaries of fair use, it unleashed the potential for unprecedented creativity that has served both the marketplace and First Amendment freedoms. The media industry, which felt threatened by the video recording technology, has also benefited immensely from the expanded global market in media products. Ironically, Betamax could not survive in the marketplace and was soon overtaken in popularity by the VCR, which is now in the last stages of becoming obsolete due to the emergence of newer products, such as interactive DVD.

FAIR USE AND UNFAIR COMPETITION

The fair use argument was tested again in another landmark case, *Harper & Row Enterprises, Inc. v. Nation Enterprises,* which concerned whether publishing copyrighted material in the name of newsworthiness and First Amendment freedom was justified.[27] Harper & Row and the Reader's Digest Association scheduled for publication in 1979 President Gerald Ford's memoir, *A Time to Heal*, dealing with the issue of President Richard Nixon's pardon. In arrangement with the publishers, *Time* magazine had contracted to serialize excerpts for a $25,000 fee, half of which was already paid, but the story was scooped by the *Nation* in its April 9, 1979, issue. The *Nation* published a paraphrased summary of the book in 2,250 words, of which only 300 to 400 words were actual quotes from the manuscript, which totaled 200,000 words. The unauthorized publication prompted *Time* to cancel the contract, thus causing financial harm to the book's publishers and triggering a lawsuit charging infringement. When the Supreme Court took up the case, it rejected the *Nation*'s argument that its newsworthy scoop based on the unpublished manuscript was a fair use and was also protected by the First Amendment. The Court also rejected the magazine's argument that the public's interest in getting a newsworthy story out as soon as possible in the precise language used by the author outweighed the author's right to control the first publication rights of his autobiography. In a 6–3 decision, the Court ruled that authors have a right to control the timing of the release of a book so that they can "develop their ideas free from expropriation," and this consideration alone "outweighs any short term 'news value' to be gained from premature publication of author's expression."

Justice Sandra Day O'Connor, writing for the majority, said that "dubbing the infringement a fair use 'news report' of the book" would "effectively destroy any expectation of copyright protection in the work of a public figure," especially if it preempted the author's first right of publication. Examining the tension between copyright and the need to scoop news in the framework of the marketplace, Justice O'Connor further observed, "In our haste to disseminate news, it should not be forgotten that the Framers intended copyright itself to be the engine of free expression. By establishing a marketable right to the use of one's expression, copyright supplies the economic incentive to create and disseminate ideas."[28] The fact that a person is a public figure does not deprive him of his right to copyright protection of his creative work. Pursuing a similar line of thinking, the U.S. Court of Appeals recognized author J. D. Salinger's right not only to the content of his unpublished letters but also to his "manner of expression" and "vividness of description," which could not be reproduced without his permission. Salinger's unauthorized biography by Ian Hamilton was based on a close paraphrasing of the content of his letters.[29]

But what if Hamilton had decided to parody the famous author? Would that have been a fair use of his letters? In *Campbell v. Acuff-Rose Music, Inc.* (1994), the U.S. Supreme Court said that parody is a fair use. In 1989, the rap group 2 Live Crew brought out their album *As Clean As They Wanna Be*, including the song "Pretty Woman," which parodied Roy Orbison and William Dees's "Oh, Pretty Woman." In reviewing the appeals court's decision that the parody had damaged the commercial value of the original song, Justice David Souter wrote for the Supreme Court's majority that some copying is essential to parody, a form of social communication that "has an obvious claim to transformative value . . . [as] it can provide social benefits, by shedding light on an earlier work, and, in the process creating a new one. Its art lies in the tension between a known original and parodic twin."[30]

MISUSING COPYRIGHT

Until the Sonny Bono Copyright Term Extension Act tilted the balance in favor of multimedia corporations, the Copyright Act of 1976 served to promote free expression and creativity through the incentive of the marketplace, nevertheless balancing it with the fair use doctrine and the First Amendment requirements of the free flow of ideas in the public interest. That's how the apparent "constitutional paradox" has been resolved through an act of balancing various public interests, though giving preference to First Amendment freedoms.[31] Many First Amendment scholars feel thrilled by a sense of absoluteness that the language of the First Amendment conveys and believe that nothing should be done to compromise that spirit. But the language of other amendments in the Bill of Right equally creates the same impression of absoluteness. The Bill of Rights is a dynamic system in which no amendment can stand alone in absolute isolation in disregard of other amendments and the Constitution as a whole.

By granting authors and creators monopoly over their work only for a limited time, after which the work enters the public domain, the Copyright Act of 1976 tried to mitigate the deleterious effect of copyright on the First Amendment. The freedom of speech and expression has also been enhanced by the fair use doctrine, allowing copying for scholarly, newsworthy, and parodic purposes. Similarly, by separating an idea from its expression, granting exclusive rights to authors over their expression but leaving facts and information in the public domain, the act aimed at encouraging creativity by allowing others to use the same set of ideas or their new synthesis for further innovation and creativity. The content-neutral policy of administering the Copyright Act by the U.S. Copyright Office has let authors write and create whatever they want, though there have been cases in the past in which copyright was denied on the grounds of blasphemy, obscenity, or defamation.[32] Since the revision of the statute in 1976, the registration of copyrights with the U.S. Copy Office is not a legal requirement, but if an infringement occurs, registration of the copyright can be helpful.

While the intent of the copyright is to stimulate creativity through the incentive of the marketplace, some famous individuals have tried to use copyright to exercise censorship on biographers attempting to probe into their personal lives. J. D. Salinger, the novelist, as mentioned earlier, succeeded in preventing biographer Ian Hamilton from quoting and closely paraphrasing the content of his unpublished letters without his consent on the ground that Salinger had the first right to the publication of his letters, which he might exercise in the future. However, Howard Hughes, the rich and famous industrialist who made his fortune in oil, movies, and aircraft manufacturing, then became a recluse in the later part of his life, failed to use copyright to impose censorship on the writer of his biography. *Look*, a pictorial magazine of the pretelevision era, published a series of three articles on some interesting aspects of Hughes's life at a time when he was not averse to the limelight. But a decade later, when he discovered that Random House had hired a writer to work on his biography, Hughes bought the copyright to *Look*'s articles to prevent the biographer from using them. Random House went ahead with the project anyway, and when the book was at the galley-proof stage, Hughes filed a copyright infringement suit against the publisher to prevent the book from being published and distributed.[33] Although the number of quotations and actual paraphrasing was minimal and would have fallen under the provision of fair use, a federal court judge ruled it as an infringement. An appeals court reversed this decision on the grounds that the biographical facts of the life of Howard

Hughes were not copyrightable and direct quotations were minimal. Since the book dealt with the life of a man who had aroused the public interest by showing his "initiative, ingenuity, determination and tireless work," the minimal copying was protected by fair use, even though the book was not a scholarly venture and was meant for the mass market. Both Salinger and Hughes wanted to protect their privacy by using copyright as a tool of censorship. Salinger succeeded in doing so because the material was unpublished, and, moreover, letters normally give the impression of intimacy. Hughes's life had already been opened up in the pages of a mass magazine, and when he tried to seal the past from the public, he failed. The Mormon Church and the Church of Scientology also used copyright to protect the absoluteness of their doctrines when challenged by hostile critics of their secretive practices.

IDEA AND EXPRESSION

As discussed earlier, the copyright statute protects a work's manner of expression but not the ideas, facts, or information it contains. It is comparatively easy to separate the two in the case of the print medium because one can use the facts and paraphrase the material; however, it is impossible to separate an idea from its expression in other media, for example, photography, music, and sculpture. The Zapruder film of President John F. Kennedy's assassination or the photographs of the My Lai massacre cannot be paraphrased effectively in words, yet their use might exceed the amount of copying allowed under the fair use doctrine.[34] When Josiah Thompson, for example, failed to get copyright permission from *Life* to use frames of the Zapruder film for his book *Six Seconds in Dallas*, he used charcoal drawings as illustrations that looked like exact copies of the frames from the film. Sued by Time, Inc., *Life*'s parent company, for infringement, a U.S. District court judge in New York City ruled that Thompson's two-sniper theory was of great public interest and that the copying was justified.[35] The important point, however, is that there was no other way of making such a significant statement, except by reproducing the film frames or approximating them as closely as possible.

Despite the Zapruder case, in which expression and idea could not be separated and First Amendment value took precedence over copyright, the traditional view holds that if an expression is not covered by fair use, unauthorized copying will constitute an infringement action. Although the original intention of copyright was to stimulate creativity by creating vested interest in the form of intellectual property, the Court always balances property rights with First Amendment newsworthiness rights. For example, when the *Miami Herald* reproduced the cover of *TV Guide* along with its own television supplement in 1971 to inform readers that they could get television programming information from the Sunday edition of the *Herald* without extra cost, *TV Guide* sued for copyright infringement. But the Federal District Court ruled that comparative advertising served the public interest by informing consumers of the availability of choices in the marketplace; therefore, the unauthorized use of the cover page of *TV Guide* was a justified fair use that advanced First Amendment values.[36]

Although facts are not copyrightable, news organizations, including television and radio stations, cannot reproduce or paraphrase information gathered by their competitors without permission under the doctrine of unfair competition. However, a news organization may use tips and leads from its competitors and do further research to write another news story. In a 1918 case, the Supreme Court established a distinction "between the utilization of tips and bodily appropriation of news matter, either in its original form or after rewriting and without independent investigation and verification."[37] A similar view, as discussed earlier, was expressed by the Supreme Court in 1985 in the *Harper & Row* case, when the Court ruled that the newsworthiness of the life of a public figure did not entitle anyone to use unauthorized material in the name of the First Amendment beyond that provided by the doctrine of fair use. Although ideas are free from copyright protection and can be recycled to create new expressive forms, they can nonetheless be misappropriated to create unfair competition, which is an anathema to the marketplace. Parody and irreverent humor are some of the best tools for dealing with the misuse of copyright to protect the absoluteness of a doctrine or a website that misuses copyright as a tool of censorship.

In 1557, Queen Mary I of England gave perpetual monopoly to London's Stationers' Company in order to exercise royal control over ideas. It might seem that we have come a long way, but the 1998 Copyright Extension Act, with its anticircumvention, copy-controlling technology, could be viewed as having the same effect on ideas, thus virtually taking us four hundred fifty years back to the future. That may not happen, however, because in the United States it is not technology that controls ideas; it is the marketplace, in tandem with the First Amendment, the bastion of core values that moves the pendulum of freedom, which will control the future of ideas. In authoritarian regimes—China, Singapore, Saudi Arabia, for example—copy-controlling technology can be, and is being, used to control ideas. It is also feared that Clickwrap and trusted systems, as an alternative to copyright, will provide further controls over ideas and creative expression, which might undermine First Amendment freedoms, but again, that will ultimately be determined by the marketplace.

ALTERNATIVES TO COPYRIGHT: CLICKWRAP

Most of us go to the Internet to get free stuff, including graphics, pictures, and, most of all, our favorite music and software programs. In our impatience to access what we are seeking quickly, we rush through the clickwrap agreement, posted as "I accept" by the content provider, without thinking of the consequences. Most of us do so with the naive hope that clickwrap agreements are not enforceable or that it will be difficult for anyone to trace us. You might wonder whatever happened to the age-old fair use privilege of the U.S. Copyright Act or the World Intellectual Property Organization, which extends the privilege to all signatory countries.

The Copyright Act gives protection to creative expressions, be they literary, musical, or artistic, provided they are tangible in form; however, as mentioned earlier, the act does not allow copyright ownership of ideas or compilations unless the compiler has added some value to the material. The white pages of a telephone directory lack the originality to qualify for copyright protection; therefore, one can copy entries with impunity. But consider the Martindale-Hubbell Law Directory, for example, which gives information about more than a million attorneys and law firms worldwide. The website is an important resource for anyone seeking authoritative information in a legal field.[38] The Martindale directory (unlike a telephone book, which is nothing but a compilation of names and addresses) is copyrighted because its publishers claim that it is unique and original. Cutting and pasting lawyers'

names and mailing them to someone for their reference, if they are looking for information about a legal specialty, should normally fall under
the fair use privilege, but if the firm's clickwrap agreement stipulates
that the copying, cutting, and pasting of content from the online directory is a violation of the agreement, are the users bound by the clickwrap
agreement or the copyright law? Can a clickwrap agreement bypass U.S.
copyright law and transcend the internationally recognized fair use privilege? The legal scenario, of course, would change if a person were a
professional information locator, someone who searches databases and
sells information to users. In this case, the searcher would be using
someone else's intellectual property, the Martindale-Hubbell Law Directory, for a commercial purpose, for which he should make a separate
contract with the company. Under the clickwrap agreement, however,
both the casual user and the professional are treated on a par, diminishing the original intent of the fair use provision of the Copyright Act.

But let's change the perspective. In order to use the Martindale-
Hubbell lawyer locator service, suppose a person had to download the
company's software, which has a built-in surveillance agent that tracks
which part of the directory the user has been surfing and how much cutting and pasting has been done. Now suppose that the person has copied
a substantial number of entries, in breach of the amount specified by the
clickwrap agreement. Martindale could take legal remedies against the
violator because the embedded intelligent agent had been "tagging"
how much had been copied. But suppose instead of suing the person for
violation of the clickwrap agreement, Martindale's intelligent agent disabled the user's browser and, in the process, deleted data files from the
hard drive, possibly incurring business losses for the person who is then
unable to meet his obligations on time. The question is whether Martindale is liable for interfering in someone's business in order to protect
its intellectual property. Electronic tags are similar to cookies, which
keep a record of one's surfing, but if tracking agents hack into a user's
database and delete or damage it, this constitutes destruction of private
property.

If you trespass onto anyone's private property, despite a clearly
posted notice reading "Trespassers Will Be Prosecuted," and you take
pictures with your video camera, what can the property owner do? He
can ask you to leave or inform the police. He can file a civil suit against
you for invasion of privacy and trespass, but he has no right to destroy

your camera or tape. Nor should he let a pit bull terrier loose on you. To correct one wrong, the law of tort does not tolerate the commission of another wrong. The law firm Martindale-Hubbell might have taken the law into its own hands by having the intelligent agent destroy someone's property without due process and a court order.

This hypothetical scenario, which I discussed in an online law course titled "Intellectual Property in Cyberspace 2000" at Harvard Law School's Berkman Center for Internet and Society, illustrates what some companies might do to protect their intellectual property. Instead of depending upon the good old copyright law, which strikes a balance between the rights of the author and the rights of the public, companies are not only thrusting clickwrap agreements on unwary visitors to their websites, but they are also developing private "trusted systems" to control information.

Writing in *Scientific American*, Mark Stefik, a principal scientist at the Xerox Palo Alto Research Center, stated that "trusted systems" are "software and hardware that enable a publisher to specify terms and conditions for digital works to control how they can be used."[39] It is feared that the copyright law, which struck a balance between the creator and the user, is being replaced by technology-based trusted systems, however without any legislative approval by Congress, which is an example of how technology sometimes trumps the law. But the real question is whether a trusted system can create trust in cyberspace and at what cost.

TRUSTED SYSTEMS

Our cities, highways, shopping malls, village commons, and homes, as well as whatever we do in physical space, are protected by technology, laws, and social norms. Social relations are based on physical structures, and cyberspace is fundamentally no different from real space, though it is perceived differently. The potential to create a digital civilization of perfect controls and regulations, according to Lawrence Lessig's terrifying catchphrase "The Code Is the Law," does exist.[40] It is feared that since computer code makers and programmers are not people's representatives but have the power to introduce controls through invisible codes, which unlike other free expressions cannot be subjected to public criticism, they might unintentionally hijack democracy and turn the United States into a corporate state. Although that may be a misplaced fear in a democratic society, in authoritarian societies the potential for misusing underlying computer codes is very great.

At the core of American society is the secularism of the free marketplace, which controls all relationships through competition. Laws are made to regulate the competitive exchange of goods and services in the public interest, but as we move toward a digital civilization, we depend increasingly upon trusted systems as an assured mode of carrying out transactions in cyberspace, especially when it comes to the distribution of intellectual goods and services. The increasing dependence by businesses on trusted systems, which protect intellectual property, should be seen in relation to the burgeoning possibilities for the free loading, instant copying, and digital distribution of computer programs, articles, music, movies, and videos. Unfortunately, there are some people who, in their messianic zeal, believe in a dangerous half-truth that all "information wants to be free."[41] If this were so, then a reward of $25 million would have gotten us some information about the whereabouts of Osama bin Laden. Information is not only not free, but sometimes you cannot even buy information. You have to extract it through intelligence gathering or some other means, whichever way you can.

Even raw information is not freely available because information gatherers and organizers have to spend time and resources to collect it. Value-added information is a priced commodity for which the user must pay so that information of high quality, be it research or a creative work, can be placed in cyberspace.

Since the existing copyright laws have not been able to maintain the fine balance between the interests of creators and the public in the digital age, when bits and bytes can be assembled, reassembled, repackaged, and shared instantly with little cost, no wonder many companies have begun to depend upon the technology of trusted systems to regain control over their intellectual property. With the help of specially developed software and hardware, trusted systems help publishers specify terms and conditions for the use of their digital property. Trusted systems can also provide greater privacy to users, thereby creating a spiral of trust that further encourages creators to put their best work in cyberspace. A person can watch a movie on a trusted system but may not make a copy of the movie for distribution unless he pays for the added use. A student can buy a one-semester right to a digital textbook and read it on a trusted system with a password, which is much cheaper than buying an expensive hard copy.

This need not mean the demise of the fair use privilege for using intellectual property or even of the freedom to copy because publishers, in their own interest or as required by the law, may offer free access to intellectual products. But some form of digital rights management based on trusted systems is essential to all cyberspace transactions, including e-commerce. Buyers and sellers in cyberspace must know explicitly what their rights and obligations are under the law, and once they agree to the terms and conditions, including price, usage, and privacy rights, transactions can take place. However, there must be a level playing field between the buyer and the seller in the digital marketplace, and that is the challenge.

The need for and level of trust may vary according to the needs of the intellectual property holder. Some publishers need to protect the absolute integrity of their intellectual work so that no one can corrupt the information or the format. If someone tries to tamper with the information, the trusted system can simply erase the information or alert the company. For example, if terrorists attempted to embed a hidden message in a file, the trusted system will not let them do it, though the system may let them view the content with or without a password and keep a trace on them.

Digital watermarking is an example of the control that a trusted system can exercise to trace the origin and use of a particular work. If someone downloads an audiovisual clip and adds it to his or her digital folder, the watermark will reveal where the copying was done. Of course, the copyright owner should know only if someone made an unauthorized copy of his work.

Trusted systems in a network establish their mutual identities through "challenge-response protocols." In simple words, a trusted computer system might ask another, How do I know that you are what you say you are? It may ask the other to solve a puzzle or a problem or to respond to a challenge in a precise, predetermined way. Trust can also be established through a digital certificate that confirms the identity of a trusted system that has been registered with an authorized company.

Stefik wrote in *Scientific American*, "Uncontrolled copying has shifted the balance in the social contract between creators and consumers

of digital works to the extent that most publishers and authors do not release their best work in digital form"; therefore, trusted systems are important because they "address the lack of control in the digital free-for-
all of the Internet. They make it possible not only for entire libraries to
go on-line but also for bookstores, newsstands, movie theaters, record
stores and other businesses that deal in wholly digital information to
make their products available. They give incentives for 24-hour access
to quality fiction, video and musical works, with immediate delivery
anywhere in the world."[42] Trusted systems for security and Open Source
software for creativity will be the two pillars of postmodern digital civilization.

OPEN SOURCE VERSUS THE MARKETPLACE

While most people think inside the intellectual property box, trying to
protect it with copyright, trademark, and contractual laws, apart from
resorting to technological means such as trusted systems, a contrarian
movement, the Open Source, or free software, movement has grown
from its initial cult status to an alternative way of doing things in the
digital age.[43] It has been argued that the Open Source free-software
movement might accomplish what the courts could not to protect the
market from domination by companies like Microsoft. If the underlying
source code of computer software were made freely accessible, it would
be easy to get rid of bugs and, through further modification, to improve
on its vulnerabilities for the benefit of all. So when people talk of free
software, they mean freedom to experiment, freedom to tinker with the
code and change it according to their needs. Software becomes communal and evolutionary rather than proprietary. If someone gets Microsoft's Windows, all he can do is use the system as it comes, even if
the company gives him a free copy. Not only is the compiled code in-
accessible, but the license also prevents any attempt to hack in. On the
other hand, an Open Source operating system such as Linux gives users
the freedom to build upon its underlying code and to develop newer
software programs, even though they might have to pay for a CD of
Linux. In this sense, Linux is free and open, though we pay for it. Microsoft Windows is proprietary, though the company might give it away
for free to schools for whatever reason. Netscape Navigator and Mi-

crosoft Internet Explorer are both free browsers, but Navigator's source code is free, and programmers have the freedom to modify it under an open license.

Paul Wallich, a supporter of the free-software movement, argues in *Scientific American*, "In some ways, it only makes sense that the Internet should run on free software: almost all its basic protocols were developed with the U.S. government funding by universities or other contractors. The Web is the brainchild of CERN, the European laboratory for particle physics near Geneva. But even after most of the Net's infrastructure has been privatized . . . development of free open software continues. . . . The logic of the intellectual marketplace ensures that only the best code and overall structure—as judged by a programmer's peers—will survive."[44] It is difficult to say whether the ultimate judges of a software program are the programming peers or the end users, but it is rather puzzling that the Netscape Web browser, despite its Open Source code, hasn't been improved by the programming community and has failed to maintain its original supremacy in the marketplace.

It must be understood that software's being available under an Open Source license, or being free, does not mean that it is in the public domain.[45] You do not need a license to use a public-domain software program, and you can do whatever you want with it. You cannot, however, take any legal action against the program's creator if there are problems. In order to protect the creators from any liability, Open Source–based software programs come without a warranty, but they are copyrighted and can be used only under a license, just like proprietary software. The difference between proprietary software and Open Source software is the degree of freedom that the latter gives under its license. The greater the freedom, the greater the possibility that users will fix problems that arise while they are using Open Source programs. Ironically, by defining Open Source software and the conditions for licensing it in exclusive terms, the Open Source movement might have unintentionally created a kind of orthodoxy of its own. Unless a license does not follow all the terms of the definition, it is not an Open Source software program. As a supporter of the movement said, "Efforts to hurt us from inside are the most dangerous. I think we'll also see more attempts to dilute the definition of Open Source to include partially free products. . . . Microsoft and others could hurt us by releasing a lot of software that's just free enough to attract users without having the full freedoms of Open

Source. It is conceivable that they could kill off developments of some categories of Open Source software by releasing a 'good enough,' 'almost-free-enough' solution."[46] This kind of restrictive attitude toward the purity of Open Source may neither be beneficial to the movement nor help it to survive in the marketplace as an alternative to proprietary systems. On the contrary, some proprietary programs may assimilate the movement's laudable features, and in that sense, it will have served its social purpose—but that need not necessarily end its existence. In a limited sense, Open Source is an alternative to copyright as an engine of creativity, but it cannot replace copyright.

Copyright is also an economic tool, and if bolstered by a socially just digital rights management system, it has the potential to offer further incentives to creativity through price differentials for intellectual products and services. At present, copyright holders have salami rights over their creations, and they can use each slice of their rights to the best economic advantage. For example, a writer may deny movie rights on a book for ten years or may allow audio rights in Europe but not in the United States. Digital rights management extends these controls electronically through a licensing system, which nevertheless poses a challenge to the fair use previsions of copyright law, a problem that legislation must address so that balance between private gains and the public good can be restored, especially in light of the recent extension of the duration of the copyright under the Sonny Bono Copyright Term Extension Act.

If the extended copyright act and copy-controlling technologies, digital rights management, and other control measures, for example, were to diminish creativity and innovation, adversely affecting the marketplace, the system of intellectual property would collapse. In time, a new balance would emerge from the resulting chaos because the marketplace thrives on order and certainty as well as creativity. Google's ambitious project to digitalize all available books in all languages and eventually all human knowledge, coupled with the public craze to YouTubize daily life, culture, and politics, is the biggest challenge that the marketplace, Congress, and the U.S. Supreme Court will face in the coming decade.

Chapter Six

Free Expression in the Digital Age

The e-mail caught my attention; it was different from the normal junk I routinely receive about second mortgages, refinancing, credit cards, debt relief, and so on. It began, "I humbly wish to seek your assistance in a matter that is very important and needs utmost trust and confidence," and it ended on a note of cautious urgency: "If for any reason you are not disposed at the moment to undertake this deal, let me know on time so as to make alternative arrangement. . . . [P]lease keep it tight secret." The phone and the fax number, included for a "heart to heart discussion," were meant to make the e-mail look genuine. The e-mailer identified himself as "a business consultant and a close confidant" of a former minister (deceased) of mines and power, the original owner of a gold company and once a member of one of the most powerful families in Johannesburg, South Africa. The new government was victimizing the family. Although the foreign and local accounts were frozen and the assets had been confiscated, the widow wanted to move US$42 million, along with an unspecified amount of gold and diamonds, out of the country. For security reasons, the family did not want to place the funds with an established institution and had authorized the e-mailer to negotiate with "a reliable and trustworthy foreigner," who could assist them to move the funds out of South Africa for safe investment. The commission for investing $42 million was 15 percent, or $6.3 million, up front and 10 percent on the annual return (after taxes) for the first five years, and thereafter to be renegotiated further. "The money is available safe with a private security company coded and lodged in with a fictitious name." I was advised to fax the person my consent, and he would let me know the steps involved in the process.

The sender's e-mail address was listed as Yahoo! New Zealand, though he said he was doing business from South Africa. Apparently, by using an e-mail anonymizer, the fraudster was hiding his identity. What distinguished the e-mail fraud, whether it came from South Africa, Nigeria, or New Zealand, from the usual spam was that the fraudster was not selling an investment opportunity for a fraudulent company but wanted to entrust me with his client's money. Believing that Americans want to make money any way they can, he must have sent the same personalized e-mail to thousands of people in the United States. Using tracer software, he would have known who opened the e-mail, and many might have done so as I did out of curiosity, especially when the message was headed "Urgent and Confidential."

I forwarded the e-mail to the State Department of Banking, Insurance, Securities, and Health Care Administration (VT) but did not receive much of a response. Law-enforcement agencies get into action only when a crime has been committed; they seldom act on mere suspicion. Even if they investigated the spam and succeeded in identifying the fraudster, they might not be able to take any action if the e-mailer lived in a foreign country.[1]

Unsolicited commercial e-mail, or spam, is a natural outgrowth of the Internet as an inexpensive medium of commerce and of the increasing use of databases as a direct marketing tool. But spam's unprecedented growth began to choke the ISPs' server capacity and Internet users' fretting about the nuisance, technological countermeasures such as bombing (returning the bulk e-mail to the sender and choking the server), filtering, and filing scores of lawsuits against bulk e-mailers, highlighted the seriousness of the problem, without offering any lasting solutions.[2] Legitimate bulk e-mailing is a serious business, as well as a form of commercial speech, but so far, unfortunately, it has been used mostly for products and services that most people find unacceptable or distasteful. To avoid detection, spammers use the names of legitimate sounding businesses and service providers in their headers to bypass spam filters. They frequently change their Internet addresses and service providers to avoid being traced by the recipients of the spam.

Bulk e-mailing can be done quickly and inexpensively with the help of special software programs, without the heavy cost of printing, handling, and postage that traditional bulk mailers incur; in fact, this would have been one of the Internet's greatest assets had there been a business model acceptable to the marketplace. But its ease of use and inexpensiveness, along with virtual anonymity through forged headers, came to be associated with "swindlers, charlatans and purveyors of sleazy merchandise."[3] Commercial unsolicited bulk e-mailers pass on the cost of their operations to Internet service providers (ISPs) by consuming their bandwidth, server space, and staff time; they also deprive legitimate users of access to their service or slow it down greatly.

SPAM AND THE FIRST AMENDMENT

Theoretically, bulk e-mailing, or spam, as a form of communication should have some constitutional protection, but the right to speak is not absolute and in certain circumstances can be regulated. In a U.S. Postal Service case regarding junk mail, the U.S. Supreme Court ruled that the "right to communicate must stop at the mailbox of an unreceptive addressee," observing that "the ancient concept 'a man's home is his castle' into which 'not even the king may enter' has not lost its vitality"; thus, the home extends into the post office box or mailbox.[4] The Telephone Consumer Protection Act (TCPA) of 1990 affords protection against junk fax mails and institutes time, place, and manner restrictions regarding telephone solicitations; for example, callers are required to identify themselves, solicit only during certain times, and take a person's name off their calling list if requested. The court balked, however, at the idea of extending TCPA to interactive communication and left it to ISPs and Congress to find a solution to spam.[5] ISPs, led by AOL, used existing statutes, such as the Computer Fraud Act (1996), the Electronic Communications Privacy Act of 1986, and the Trademarks and Service Mark Infringement Act, to obtain restraining orders and judgments against spammers.[6]

In the case of *AOL v. Cyber Promotions,* the question before the court was whether AOL had created a public forum, obligating the service provider to open access to the public on a nondiscriminatory basis; if it had, did AOL's retaliatory measures, such as bombing, amount to a state action? U.S. District Court Judge J. Weiner ruled that AOL "does not stand in the shoes of state" and has not been acting on behalf of the government in providing a public service; therefore, the company was under no obligation to give access to Cyber Promotions for communicating with the client base of America Online.[7] Had AOL been the only means of communication with the people, that is, if the company had a monopoly, the denial of access would have created a different kind of problem. But Cyber Promotions had other means of delivering advertising messages to AOL customers, including the U.S. mail and the news media. The First Amendment issue in the absence of any state action does not touch private parties, unless there is a situation where a private party's action might amount to an unfair trade practice. For example, consider a hypothetical case of a small town, where the only outlet for a business to advertise its product and services is the town's only media outlet, its advertising-based daily newspaper. If the newspaper, for no legitimate reasons of policy, refuses to sell advertising space to the business, a justifiable case could be made. Although the newspaper is protected by the First Amendment only from government restraint on speech, its refusal to sell the business space, in other words, to allow it to use the forum to deliver its commercial message, could amount to an unfair business practice and therefore be actionable if no alternative venue for commercial speech was available.[8] Because AOL did not exercise monopoly rights and alternative avenues of access were available to Cyber Promotions, AOL had the right to protect its subscribers from unsolicited junk mail, the court ruled.[9]

Successful legal actions taken so far under various federal and state fraud statutes, including misappropriation, misrepresentation, unauthorized access, damage and destruction of stored communications, trade and service mark infringements, false description, and false designation, have not finally settled the constitutional issue of unsolicited e-mail. The courts have been sympathetic to ISPs, partly because in most cases spam involves pornography and fraudulent schemes. How the courts will react if a group obtains an AOL member list and spams messages regarding AIDS or abortion, along with commercial messages, remains to be seen. Under *Central Hudson Gas & Electric Corporation v. Public Service Commission of New York* (1980), the U.S. Supreme Court extended First Amendment protections to include commercial speech. The right to advertise cannot be abridged by the government unless it meets a four-part test. Justice Lewis F. Powell Jr., writing for the Court, stated,

At the outset, we must determine whether the expression is protected by the First Amendment. For commercial speech to come within that provision, it at least must concern a lawful activity and not be misleading. Next, we ask whether the asserted government interest is substantial. If both inquiries yield positive answers, we must determine whether the regulation directly advances the government interest asserted, and whether it is not more extensive than is necessary to serve the interest.[10]

Although commercial speech, which also covers spam or unsolicited commercial e-mail, has received conditional constitutional protection, the government cannot act arbitrarily and has to follow due process before it can take any action. Federal and state lawmakers have been subjected to contrary pressures; consequently, the legislative remedies they have proposed have raised controversies either because they were deemed too ineffective to solve the spam problem, were excessively market oriented, or were so overbroad that they would have adversely affected the flow of information in the digital society. On top of it, there is a pervasive fear of the government's stepping in and regulating the Internet. The Direct Marketing Association (DMA) supports direct e-mail and electronic-database marketing and is opposed to any legislation regarding unsolicited commercial e-mail. DMA supports self-regulation and allows consumers who do not want to receive unsolicited messages to remove their telephone numbers and addresses from its preference lists. But it does not maintain ISPs' customer lists, which, being proprietary in nature, are protected for competitive reasons. DMA cannot protect consumers from spam.

Most ISPs would rather look to effective filtering solutions to control the nuisance of spam.[11] For example, an opt-in filtering system can be used to allow only those e-mail messages that are expected by recipients according to a predetermined list. The system will automatically reject any mail that does not meet the criteria. An opt-out system will receive all e-mail but screen out any mail from addresses it has been specially programmed to reject. The users themselves can opt in and out by turning the system on or off, or they can ask their service providers to do the necessary filtering for them. Although filtering is a better option than the total banning of unsolicited commercial e-mail, the serendipity of an encounter with the unknown, one of the greatest adventures of living in the postmodern digital age, is lost when e-bots

mess with the free flow of information. Spamming is essentially a marketing problem, and eventually the marketplace will refine it for its purposes or it will go the way of snake-oil peddling.

FACTOIDS AND HALF-TRUTHS

Internet spam is not the only problem in the digital age. A growing culture of factoids, half-truths, and infotainment presented as news tends
to prevent serious social issues from being thrashed out in the marketplace of ideas. Network television news in the United States is fed to the
audience in a series of small tablets, each containing a video clip and a
standby talking-head correspondent and concluded with commentary
by an anchorman. To deal with viewers presumed to have an attention
deficit and to be restless with the remote control, and to bring them back
to the news after commercials that interrupt the news flow, the anchorman shows them a teaser now and then. Creating anticipation and a
sense of drama in the segmented newscast has become an essential part
of the news strategy, even though it sometimes distorts reality.

But television news as infotainment has become a norm in the United
States. Profit, rather than news credibility, is the driving force behind
the media's news strategy—news, as a profitable commodity, must be
exciting and entertaining. There was a time when the American media
gave the people all that was newsworthy, which a handful of serious
newspaper organizations still do today.

When the culture of spam, factoids, and infotainment extends to
news, what happens to serious news? Dr. Jack Kevorkian had to shock
Americans out of their entertainment stupor by videotaping the death of
a patient; Kevorkian persuaded *60 Minutes* to air the videotaped killing
on November 22, 1998, highlighting the problem of physician-assisted
suicide. For CBS, the mercy-killing episode, which promised to garner
a huge television audience during the November sweeps month, which
determines advertising rates, was a profitable scoop. In 1993, NBC's
Dateline staged the explosion of a GM truck in order to beef up its investigation and support its predetermined conclusion that the truck was
unsafe to drive. The complaint against the truck's safety was based on
sufficient anecdotal evidence, but the newsmagazine wanted to demonstrate to the audience how the explosion could occur by actually faking

it to create dramatic impact. In search of the sensational, the NBC news program put its credibility on the line, which did not help its ratings and advertising revenue for the year. Declining credibility is a serious problem for the American media in the age of twenty-four-hour news cycles and the rush to deliver news as it happens.[12] But by blending news with theatrics or sensationalizing news to hold the audience, television networks are turning news into cultural spam, which is no different from online spam. Since the American media dominate the world via satellite communication, they spread the same culture of spam, half-truths, and factoids around the globe. Al Jazeera, the Arab news network, which claims to follow "the journalistic values of honesty, courage, fairness, balance, independence, credibility and diversity," though it actually seems to be working according to the motto "if it bleeds, it leads," was started in 2003 in response to the culture of factoids and infotainment spread by the American global media.

Like surround sound and the wraparound images of Cinerama, we are enveloped by news day and night. We used to say, no news is good news; now we say, no news is news unless it is breaking news. Consequently, in a twenty-four-hour news cycle, we do not feel the thrill of the news, such as "man bites dog," a deviation from the normal. The daily news has become soap operatic, endless dovetailing stories punctuated by newsbreaks. The twenty-four-hour news cycle, which began with CNN in 1980 and reached a high point with its seventeen-hour, nonstop, live broadcast from Baghdad during the Gulf War in 1991, has found many followers, such as MSNBC and Fox, but it has killed the sense of the occasion or urgency that news is supposed to generate. When Edward R. Murrow reported from London during World War II, America listened in awe. He became a legend with his documentary *Harvest of Shame* and the newsmagazine *As You See It*, which helped end the political terrorism of Senator Joseph McCarthy. Walter Cronkite was the last news anchorman whose broadcast sounded important enough to create trust in what he said about the way the world was, whether it was the death of President John F. Kennedy or the landing of the Apollo on the Moon. Today's news anchormen sound more like news peddlers, dressing up news for sale, purveyors of news on a perpetually revolving infotainment conveyor belt, than like authoritative editorial voices, honest and bold enough to say, This is how the world has been lately. These newspeople, with all the attitude of actors and superstars, multi-million-dollar celebrities, are not the kind of news hounds and serious analysts who can explain the world to us. Nevertheless, not all is lost. Constant hammering of information by the repetitive news cycle might in some cases create social awareness about some social problems and lead to individual and collective action.

BENEFITS OF THE 24/7 NEWS CYCLE

Despite the ecology of spam and infotainment, if the viewers are savvy enough to discriminate what is worth watching, news around the clock can sometimes be informative and enlightening. It seemed that Americans were better informed about the legal system because of the year-long television trial of O. J. Simpson, charged with murdering his ex-wife, Nicole Brown, and her boyfriend in 1994. Blow by blow, day after day, television newscasts laid open the complexities of the American criminal justice system, including how trial by a jury of one's peers works. The trial was an eye-opener. Many people did not feel that justice was served and felt unhappy about the jury verdict of acquittal in the criminal trial, but they felt vindicated by the subsequent civil jury verdict holding Simpson liable for the double murder and the award of damages to the bereaved families. As a consequence of the interest in the trial, CNN began to broadcast a daily show, *The Burden of Proof*, which subsequently discussed the ongoing legal aspects of the impeachment inquiry against President Bill Clinton and other important legal cases. It might seem, though empirical evidence is not available, that the American people today have better awareness about the Bill of Rights and the workings of the Constitution than ever before, thanks to Court TV and its legal experts, who explain what is happening in the world's most litigious society.

No less important is how the news about medicine and health care gets embedded into the news cycle. Today, there is greater awareness of AIDS and the associated risky behaviors that cause the disease, and that might explain why the incidence of new cases has dropped in the United States; perhaps the same would happen in other countries if they had a twenty-four-hour news cycle and fewer taboos about talking about sex. Frequent information about breast cancer piggybacking on the news as serial reports or minidocumentaries alerts women to the need to have periodic checks and consequently affects policy decisions and stirs compassion for raising financial resources for research. Because of the constant drumbeat about food-safety issues in the daily news, we are better informed about what we should do to protect ourselves from salmonella and E. coli. Some experts say that recurrent television news about murder and mayhem desensitizes us to violence and dulls our compassion. But there is another side to the question. Reportage of crime in television motivates politicians to take action, as the Clinton administration did by providing extra funds to hire more cops and build more jails, which led to a decline in violent crime.

The understanding of most people about social problems is limited by the quality of the news media stories they are exposed to. A serious and responsible twenty-four-hour news cycle, not the kind that is there to fill airtime, has the potential to lift us from ignorance to awareness by giving us sufficient information, thus making us ready for political action. News around the clock cannot make us experts, but it can enable us to engage in a dialogue with experts. The assumption that some knowledge about a problem that concerns people deeply may instigate them to seek expert advice has led pharmaceutical companies to advertise their prescription drugs on television, urging people to talk to their physicians; this has no doubt pushed up the cost of prescription drugs, but it has also raised awareness about health issues. In this oversaturated environment of spam, factoids, and news gluts, Internet search engines and portals have come to play an enormous role in enabling people to seek trustworthy news.

WHEN ACCESS IS DENIED

Corporations are fighting for access to consumers by luring them with news, sports, information, entertainment, or whatever else it takes. If they cannot reach enough of them, or the right ones, however, their survival in the marketplace becomes problematic.

A financial dispute between two multimedia giants, Disney and Time Warner, came to such an impasse in 2000 that 3.5 million households lost access to their favorite news and entertainment programs, including a most popular game show at that time, *Who Wants to Be a Millionaire*.[13] Disney, which owns ABC, is one of the biggest entertainment companies with a global reach. Time Warner, the parent company of HBO, has one of the largest cable systems and controls access to millions of viewers in the United States. Its merger with AOL and CNN, and the convergence of technologies to create a broadband multimedia superhighway, held the potential to create unprecedented customer access. The business-to-business quarrel between the two companies was about the level of financial compensation that Time Warner must pay Disney for retransmitting its programs to its cable customers. Before the 1992 Cable Act, the retransmission of local and network broadcast programs was free for cable companies, as well as a requirement under the "must carry rule." But after the 1992 act, the financial rules changed, and cable companies had to pay for the retransmission of programs.

May is a sweeps month, one of the four important periods during which advertising rates are determined for networks. One point in ratings represents more than a million households and that means a lot of advertising revenue, which prompts television networks to put their best shows on during the sweeps months to earn the highest ratings. Because of the dispute, Time Warner cut Disney-ABC off from access to its viewers in several major markets, including New York and Houston. Thus, a simple business dispute became a political question. Who owns access to viewers? Can a media company with a local and regional monopoly pull the plug and refuse access to a content provider, when everything from entertainment to e-commerce has become totally dependent on electronic and digital communication? Denial of access to an audience in certain circumstances might be construed as a form of censorship seldom experienced before; it's also an anticompetitive and unfair market practice. Freedom of speech is tied up with access to the free marketplace, and there is lot of money to be made in creating and controlling new avenues of access to the public. AOL-Time Warner's merger held the possibility of generating new opportunities for interactive access, such as commercials with hyperlinks that would allow viewers to watch television and order goods online. Control over access by the newly merged company AOL-Time Warner created fear in the minds of Disney's executives, since Disney had content but insufficient control over access. Public protests and threats of congressional hearings defused the crisis between Disney and Time Warner, but the crucial issue of access remained unresolved until recently, with the possibility of content's being distributed over broadband Internet via Joost and YouTube. If access is the key to success in business as well as politics in the digital age, how should it be made abundant and free so that no corporation can cut the public off from the free flow of information necessary for the healthy functioning of the marketplace of goods and ideas? The words of Supreme Court Justice Oliver Wendell Holmes Jr., dissenting in a case over the distribution of antiwar leaflets during World War I, are still very relevant today:

> But when men have realized that time has upset many fighting faiths, they may come to believe even more than they believe the very foundations of their own conduct that the ultimate good desired is better reached by free trade in ideas—that the best test of truth is the power of the thought to get itself accepted in the competition of the market; and that truth is the only ground upon which their wishes safely can be carried out.[14]

It is disputable whether the United States or any other society can ever have a true marketplace of ideas, especially when a few big media conglomerates or governments control the mass media, including Internet portals.[15] But even to have some openness of ideas in the marketplace, for commercial, cultural, and political expression, access to privatized public spaces is crucial. The U.S. Constitution restrains the government from interfering with free speech, but there is no provision to stop private companies from denying access to the media resources they control. Nor is there any stipulation that the government must promote free speech through affirmative action, except in the copyright clause, according to which the government is expected to create intellectual and creative property rights in order "To promote the Progress of Science and useful Arts." But private media companies can deny access to their media outlets for competitive reasons, or they can make it prohibitively expensive for individuals.

Denial of access to the media, whatever form it takes, is the worst form of censorship. As law professor Jerome A. Barron wrote in the era of civil rights and the Vietnam conflict, "Freedom of the press must be something more than a guarantee of the property rights of media powers."[16] The U.S. Supreme Court rejected the idea of a mandatory access rule in a landmark case in 1974. Pat Tornillo, a Florida schoolteacher who was a candidate for the Florida House of Representatives, was criticized in an editorial published in the *Miami Herald* for his role as the executive director of the Classroom Teachers' Association. A Florida law passed in 1913 required that if a newspaper attacked the character or official record of a person seeking public office, the right to reply must be published free of charge. Taking advantage of the state law, Tornillo demanded that the *Miami Herald* publish his response verbatim, but the newspaper refused to oblige him. When the U.S. Supreme Court finally accepted the case for review, it ruled that the Florida right-to-reply statute was unconstitutional. Chief Justice Warren E. Burger, writing for the unanimous court, said, "A responsible press is an undoubtedly desirable goal, but press responsibility is not mandated by the Constitution and like many other virtues it cannot be legislated." He mentioned four major problems with mandatory access to the press. Mandatory access is no less injurious to press freedom than a state action, which is prohibited by the First Amendment. The cost of publishing the right-to-reply is an unnecessary financial burden on the press. In order to avoid the cost of publishing free right-to-reply statements, the press might avoid controversial topics, which will chill free speech. Mandatory right-to-reply trumps the editorial control and judgment of newspapers, thus adversely affecting the freedom of the press because, as he said, "A newspaper is more than a passive receptacle or conduit for news, comment, and advertising."[17]

FEAR OF FREE EXPRESSION

In July 1996, the French government tried to keep a tight lid on information about President François Mitterrand's cancerous prostate, much in the tradition of the Soviet Union, which kept the health of its top Communist leaders a closely held secret until everyone was sure that dead men could do no harm. But Dr. Claude Gubler, Mitterrand's physician, decided to tell the story to the public and authored *Le Grand Secret*, describing the president's suffering in graphic and gloomy detail.[18] A French judge banned the book, presumably to spare the president's family the pain and humiliation of public exposure, which was rather strange for a country where freedom of expression has been long cherished. But *Le Grand Secret* soon appeared on websites in Canada, Britain, the United States, and many other countries, with the complete text and hideous pictures of the much admired and controversial politician. Through hypertextual links, readers also learned about the president's checkered past, romantic and unsavory, for example, about his twenty-year-long affair with his mistress, about their daughter, and about his association with the infamous Vichy government during the Nazi era. Dr. Gubler was more interested in letting the world know about the final days of the French president than in copyright violation when the book was uploaded onto the Internet, probably not without his implicit consent. Online book sharing, even if it is purported to beat censorship, has seriously begun to worry publishers of trade books with hot-selling titles, exemplifying the clash between free speech and the marketplace.[19] At the heart of every culture, there is censorship, denial of access to ideas that might be too disturbing for people in power or that threaten the system's equilibrium.

In 1663, just after the Restoration, which brought Charles II to the English throne, William Twyn was sentenced to be hanged for printing a book supporting the right of revolution and imagining the king's death. According to historian Leonard Levy, Twyn was "cut down while still alive, and emasculated, disemboweled, quartered, and beheaded— the standard punishment for treason."[20] Europe, like the rest of the world, has a long tradition of suppressing thought and expression when ideas do not conform to the official truth or make people in power uncomfortable. Ultimately, no harm came to Dr. Gubler, and after the curiosity and scandal had died down, the book finally found its resting place on bookshelves as generally happens with kiss-and-tell, quickie books written by insiders.

Though we might want to celebrate the triumph of technology in finally beating censorship, it may be rather premature to do so because the very technology that helps combat censorship also enables authoritarian rulers to use filtering technology, along with social and market controls, to impose censorship on their people, as China has done in silent collaboration with some American media companies, namely, Google, Yahoo!, Microsoft, and Cisco, among others. Apart from the suppression of ideas with technology, sheer fear of consequences can cripple free expression. It is difficult to surmise the effect that the fatwa against Salman Rushdie, author of *The Satanic Verses*, has had on other writers, who might have succumbed to self-censorship in writing about Islamic society.[21] Rushdie has not written another book critical of Islam, and one can see the steady decline in his creative imagination after the fatwa. One might call it the nonviolent murder of a promising writer.

Although attempts are made in almost every country to censor free speech and expression, including banning books, it has become increasingly difficult to suppress information for long in the digital age. What disappears from the bookshelf reappears on the Internet with a louder voice and in many languages, due in part to fast and affordable scanners but mainly to the fact that well-organized anticensorship groups are determined to defend the people's right to know. Even in the United States, long regarded as the bastion of freedom and liberal democracy, every year attempts are made to censor books; the books are not banned by the government, however, but by local authorities, mostly school and library boards, trying to protect children from exposure to racism, obscenity, or the threatening alternative lifestyles portrayed in some books, for example, *Heather Has Two Mommies* by Leslea Newman and *Daddy's Roommate* by Michael Willhoite. The U.S. Supreme Court in *Miller v. California* (1973) wrestled with the definition of obscenity but passed the responsibility on to local authorities, who were to use the test of "contemporary community standards."[22] But the digital revolution has made nonsense of the concept of "contemporary community standards," not so much because localism has become irrelevant but because inquisitive and restless young minds have the tools to break through filters and surf wherever they like. The Internet is becoming widely available in school and local libraries, and websites dedicated to banned books with complete digital texts can be accessed with a keystroke. But in an open society, for every social action there can be an equally vehement reaction, which occurs in the form of a backlash, as one can see in the clamor for filtering software to protect children.

Many school libraries have kept Mark Twain's *Huckleberry Finn* off their shelves because of the writer's use of the word "nigger." Today, the word "negro" is politically incorrect and has been stricken from the American lexicon; "nigger" is now obliquely referred to as the "N word," as was done during the O. J. Simpson trial. The list of banned books is long, but at one time or another, at some place in the United States, books, some of them considered classics, for example, *Ulysses* by James Joyce, *Fanny Hill* by John Cleland, and *Lysistrata* by Aristophanes, were banned under the Comstock Law of 1873. Walt Whitman's *Leaves of Grass* was frequently banned for homosexual references (for example, "I am He that Aches with Love/I am He that aches with amorous love"). In 1996, Merrimack schools in New Hampshire banned Shakespeare's *Twelfth Night* under the school board's "prohibition of alternative lifestyle instruction" guidelines.[23]

Digital technology liberates our minds by giving us a wealth of information from all over the world, but it can also become our worst censor. In their fight against the technological censor, it is fascinating how digital censor breakers and hackers of the cyber age are working to keep the book, one of the greatest inventions of mankind, as free as it was when Johannes Gutenberg invented the movable-type printing press and sowed the seeds of Europe's Reformation, which was nothing but the breakup of the power of the Catholic Church and consequent liberation of the European mind from the tyranny of religious infallibility and absolutism. With the freedom of the Internet, it seems humankind may be poised for another revolution and a period of turmoil.

CHECKERED HISTORY OF FREE SPEECH

The history of free speech is at best an uneven one. The Athenian democracy, which lasted for about two hundred years (beginning in 508 BC), soon discovered the limits of free expression in the trial and state-ordered suicide of Socrates (399 BC). Plato's ideal republic was based on a system of indoctrination and thought control. Ancient Rome experimented with some form of citizen lawmaking and popular rule for more than four hundred years (449–44 BC), until imperial Caesars seized all power.[24]

For more than fifteen hundred years, Europe lived under the paradigm of the divine rights of kings, which depended upon the brute force of the militia, control of trade and commerce, and alliance with the Catholic Church, which preached obedience as a most important human value. In this powerful alliance between king and church, the source of truth was revelation and tradition rather than freedom of thought, which made free speech not only unnecessary but a threat. Censorship became an indispensable tool for preventing unacceptable truths from emerging, truths that did not conform to the received tradition. The Roman Catholic Church developed the *Index Librorum Prohibitorum*, which contained a list of books prohibited because they did not subscribe to the church's orthodoxy and its monopoly on truth.[25] Since Galileo's

1633 discovery of the truth about planetary motion, for example, did not accord with the church's doctrine about the Earth's position in the solar system, he was forced to recant.[26] The trials of Joan of Arc of France in 1431 and Sir Thomas More in England in 1533 were based on the fact that they challenged the prevalent official view.[27]

The change began to occur in the fifteenth century when the printing press, first developed by Gutenberg in Germany between 1453 and 1455, let loose forces that eventually led to the Protestant Reformation, the Renaissance, and the Enlightenment. With the spread of printing technology through the rest of Europe, the church lost its monopoly over the dissemination of knowledge, which it had exercised by supervising the production of book manuscripts in monasteries. Religious dissenters like Martin Luther, a Catholic priest appalled by widespread corruption in the church, urged people to go to the source, the Bible, rather than depend on the papacy. While the church was being questioned by both the clergy and the laity, explorers, artists, musicians, painters, astronomers, and philosophers were challenging the authority of the king. Europe was breaking away from the orthodoxy of the church and the autocracy of the monarchy, institutions whose authority was based on the revealed truth and the king's divine right to rule. In England, for example, after the beheading of Charles I in 1649 and the Restoration following the Puritan Interregnum, the monarchy was compelled to share power with the parliament. The power-sharing arrangement did not expand freedom of speech in any measure immediately but John Milton's treatise *Areopagitica* questioned divine revelation as the source of truth. For *Areopagitica*, Milton, a devout Puritan and the author of *Paradise Lost*, did not seek inspiration from the Bible but from the supreme tribunal of Athens, the Areopagus. If truth can emerge out of unfettered debate in the marketplace of ideas, then censorship in any form is a handicap.

While Milton had faith that in the contest of ideas truth would emerge victorious against falsehood, John Stuart Mill, the English philosopher, said that contest alone was not necessarily enough between truth and falsehood. In his essay "On Liberty," he argues that the voice of the minority may also contain truth and its silencing will result in "robbing the human race: posterity as well as the existing generation; those who dissent from the opinion, still more than those who hold it. If the opinion is right, they are deprived of the opportunity of exchanging error for truth; if wrong, they lose, what is almost as great a benefit, the clear perception and livelier impression of truth, by its collision with error."[28]

History teaches us that truth does not necessarily lie with majority opinion, which can be manipulated and controlled through propaganda. Dissent must be given a chance for full expression because it might emerge as acceptable truth in the future. It must be noted, however, that despite the place of dissent in the marketplace of ideas, freedom of speech may not settle controversial issues, and in exceptional circumstances, truth might have to be imposed by an authority, as happened in the U.S. presidential election of 2000, when the U.S. Supreme Court stopped the Florida vote count and George W. Bush was declared the winner by default. The Supreme Court's decision became the truth, though we will never know what the actual truth was. In many ways, the U.S. Supreme Court has taken on the moral authority of the medieval Catholic Church, and its decisions are regarded as beyond challenge. Human beings need boundaries to order their lives. The pendulum of freedom moves over the dome of attraction, the bastion of core values that bonds a society and defines the space for freedom. In the United States, the Supreme Court holds the unique position of finally determining what is truth and how much freedom and privacy is good for the people.

In 1734, truth became the touchstone for a decision regarding a case of seditious libel; the seditious libel law had been an instrument of controlling dissent in England and had been extended to the American colonies. John Peter Zenger, publisher of the *New York Weekly Journal,* is memorialized as a torchbearer of freedom because he defied the seditious libel law and published articles critical of New York governor William Cosby, accusing him of corruption, oppression, and incompetence. Under the sedition law of the time, any statement, regardless of how true it was, if it tended to diminish the people's faith and trust in their government, was culpable. Under the principle "the greater the truth, the greater the libel," the judge instructed the jury to establish the facts of whether the accusations were likely to reduce the people's faith in the government. But under the persuasive advocacy of Zenger's lawyer, Andrew Hamilton, the New York jury ignored the judge's instructions, considered his statements' truth as his defense, and found Zenger not guilty. Although the Zenger case did not end the seditious libel law, it established a precedent that truth could be a defense against charges of libel in some circumstances. More importantly, however, the Zenger case is regarded as a milestone for the freedom of the press because it showed that the press can work as a check on government.

The idea of limiting government's power had become widely prevalent since John Locke's *Two Treatises of Government*; Locke's thinking had a definitive influence on the founders of the American Republic. Developing further the social contract theory of the political philosopher Thomas Hobbes (1588–1679), Locke argued that the government's power should be limited and that government should be held accountable for the way it exercised that power. To safeguard the freedom of the people, Locke propounded a constitutional structure that clearly defines the obligations and powers of the government and allows the people to retain their natural rights, which they might have enjoyed in the state of nature, as Hobbes had theorized. This could be best achieved through a tripartite structure of government: an elected legislature that makes laws, an executive that carries out the laws, and a judicial system that administers justice and protects the rights of the people. Broadly based on Locke's ideas of a constitutional government, the American founding fathers adopted between 1789 and 1791 a system of government divided among three coequals in power, limited by the Bill of Rights, thus culminating the long journey from the divine right of kings and the revealed truth of the church and crossing a threshold where free speech presented endless possibilities and endless dangers. In this journey toward freedom of speech and the press, the marketplace created by printing technology played a role no less significant than that of poets and philosophers who provided the rationale and rhetoric for freedom. The emergence of printing and the book trade as a market force brought about a radical change in thinking about licensing and censorship:

> Economic goals and profit were the central interest of the growing numbers of these tradesmen in the late Seventeenth Century; hedged and bound by the Regulation of Printing Act, cut out of the privileges still granted guild printers of the Stationers Company, they sought relief from Parliament. Unsuccessful in 1692, they continued pressing, and with help from people of power including philosopher John Locke, won their way in 1695. The House of Commons, offering a long list of reasons for its refusal to renew the Printing Act, *focused on the restraints of the trades as the main factor* [italics added], saying nothing about the principles of freedom of the press.[29]

It requires a stretch of one's historical imagination to believe that the Magna Carta, signed by King John in 1215, was the beginning of free speech in England. By limiting the power of the king, however, the English upper classes no doubt sowed the seeds of a future civil strife that resulted, centuries later, in the signing of the Petition of Right (1628), the Habeas Corpus Act (1679), and the Bill of Rights (1689), thus eventually preparing colonial America for the Declaration of Independence in 1776, which ultimately severed the colonies' umbilical cord with the motherland after more than five years of war. The founding fathers of the American Revolution crafted a constitution with a strong rider, the Bill of Rights (1791), a slate of citizens' rights so broad and vague that every succeeding generation of Americans has had the freedom to interpret its meaning according to its needs. The First Amendment to the Constitution, prohibiting certain actions by the government, has particularly been the rock on which civil liberties, including freedom of speech and expression, have been continuously built, elaborated, and defended:

> Congress shall make no law respecting an establishment of religion, or prohibiting the free exercise thereof; or abridging the freedom of speech, or the press; or the right of the people peaceably to assemble, and to petition the government for a redress of grievances.[30]

Just as the founders drew heavily upon Locke's notion that the source of the government's power is the people themselves, they were also aware of the legal commentaries of the English jurist William Blackstone, who, like John Milton, condemned government censorship of the press, but argued that it was necessary for the government to punish "improper, mischievous, or illegal" expression. Blackstone commented,

> The liberty of the press is indeed essential to the nature of a free state; but this consists of laying no previous restraint upon publications, and not in freedom from censure for criminal matter when published. Every freeman has an undoubted right to lay what sentiments he pleases before the public: to forbid this is to destroy the freedom of the press: but if he publishes what is improper, mischievous, or illegal, he must take the consequences of his own temerity.[31]

SEDITION AND FREE SPEECH

Despite the First Amendment, the history of the freedom of expression in the United States has been a continuous struggle between prior restraint of expression that nips freedom in the bud and postpublication punishment of expression that's as chilling and fearsome as prepublication restraint. The first major challenge to the freedom of speech came within the first few years of the establishment of the new republic, when Congress passed the Sedition Act of 1798. The intent of the act was to control the passions that war between England and revolutionary France aroused in the American people, and the act made it a crime to indulge in harmful criticism of President John Adams and his government. The Sedition Act also symbolized the painful emergence of the two-party system in the United States and the attempt of one party to undercut the influence of the other. John Adams's Federalist Party sympathized with England; nonetheless, the government wanted to keep the United States out of the European war. On the other hand, the Republican Party of Thomas Jefferson, remembering the role the French played in the American War of Independence, was sympathetic to the French. Wars in Europe have always created strong partisan feelings in the American people, primarily because most Americans trace their origins to different regions of Europe. In the struggle between the Federalists and the Republicans over how to deal with the European crisis and domestic partisan politics, freedom of expression became a casualty. The newspapers of the time were blatantly partisan, and obviously the victims of the Sedition Act were primarily the editors of the Republican newspapers who criticized John Adams's handling of European affairs.

Although the Bill of Rights was created as a supralaw, a law above the law of the Constitution, the American people were not fully aware of the force of the law. Nor did the U.S. Supreme Court play the same role in the lives of the people that it does today, that is, protecting the people from the excesses of the government, thus playing its constitutional role as a coequal in power and the final arbiter of what is freedom and how much a person needs. Justice Samuel Chase of the U.S. Supreme Court, acting as a trial judge, in his decision against editor Thomas Cooper for seditious libel, wrote that "licentiousness of the press" could destroy a government.[32] Because of its sunset clause, the Sedition Act expired in 1801, coinciding with the end of John Adams's presidency. And Thomas Jefferson, after he assumed the presidency, pardoned all of the journalists who had been convicted under the law.

The important point is that when all three branches of the government are united on an issue, there is a high probability that the people's civil liberties might shrink. The greater the distance between the three branches of government, the greater is the room for freedom of speech and expression to flourish. Only when free speech cases, vociferously supported by prominent legal scholars and philosophers, began to be accepted by the U.S. Supreme Court in the early part of the twentieth century did the First Amendment come alive from its dead slumber and become a dominant force in American society.

Gradually, it began to dawn on society, as John Stuart Mill stated in his essay "On Liberty" and as legal scholars later discussed further, that arriving at rational decisions through open and free discussion was the most desirable method of resolving differences. Freedom of speech began to be perceived as an instrument for bringing all kinds of opinions, be they of a plurality, majority, or minority, into an open forum for the greater good and stability of society. The Supreme Court's decisions reflect the mode of consciousness of the people and their network of assumptions at a particular time in history, but those assumptions are articulated by legal scholars in their writings or in the briefs they present to the Court in landmark cases. Because of the breadth and vagueness of the language of the speech and press clause of the First Amendment and its relation to regulating the economic, political, and social lives of the people, myriad interpretations have become possible. Since no one is sure what the founders meant precisely by "Congress shall make no law," it has become necessary for each generation to define freedom once again.

But legal scholars and philosophers have also recognized that free speech can cause harm to individuals and society. The history of free expression in the United States is nothing but a struggle to define what the society considers to be "harmful" speech and how best to deal with it. In other words, it amounts to interpreting the First Amendment's absolute command not to interfere with free speech while at the same time setting boundaries on free expression so that the government can meet its constitutional obligations and the people can equally enjoy other fundamental rights enshrined in the Bill of Rights. If it is a self-evident truth "that all Men are created equal, that they are endowed by their

Creator with certain inalienable Rights, that among these are Life, Liberty, and the Pursuit of happiness," then the government must have the power, as the declaration says, to "establish justice, insure domestic tranquility," and "secure the blessings of liberty to ourselves and our prosperity."[33]

If free speech and expression come into conflict with the constitutional obligations of the government, what can be done? It is worth noting that while the First Amendment's commanding language forbidding government to abridge freedom of speech and expression seems absolute, the Fifth Amendment (as well as the Fourteenth Amendment) includes a clause under which a person cannot be deprived of his liberty except through due process. In this sense, the Bill of Rights should be considered a total system of fundamental rights that remains in dynamic equilibrium. At any given time, beginning in the twentieth century, there has been tension both between First Amendment freedoms and other fundamental rights and between the First Amendment and the government.

HOW THE SUPREME COURT RESCUED THE FIRST AMENDMENT

During much of the nineteenth century, after the Sedition Act expired in 1801, First Amendment freedoms were not seriously contested in the court. It was a century of civil turmoil, affirmation, invention, and expansion. The United States underwent serious internal convulsions that tested its national integrity and the values on which it was founded. During the Civil War, censorship of news from battle zones, both in the North and the South, was common and was accepted without much protest. After the Civil War and during its epic expansion into the West, the United States emerged as a networked nation through the development of the railroad, the telegraph, the telephone, and later the radio. The emerging communications technologies of the late nineteenth century and the expanded market they created for information services began to impact the gathering and distribution of news, thus pushing First Amendment freedoms to the Supreme Court's doorstep for interpretation at the beginning of the twentieth century.

Just as the fear of terrorism led to the enactment of the Patriot Act in 2001, the declaration of war against Germany in 1917 persuaded Congress to enact hastily the Espionage Act of 1918. The position of Arab and Muslim Americans in the post-9/11 period is not very different from that of German Americans during World Wars I and II and of Japanese Americans in World War II (the latter were even put into internment camps). The Espionage and Sedition acts aimed to criminalize any expression of opposition to the government's war efforts. The Supreme Court's decision in a 1919 case, *Schenk v. United States*, that Congress was constitutionally empowered to punish speech that prevented its efforts to wage war, was the beginning of its involvement in the interpretation of the First Amendment, which in time began to transform the United States into an open society the likes of which the world had never seen.[34] The Court's rulings dealing with various landmark cases regarding the First Amendment have provided philosophical reasoning as to why the freedom of speech lies at the heart of a democratic society wedded to the free marketplace. The case-to-case broadening of the freedom of speech in the twentieth century has not only strengthened democracy through increased public participation in the affairs of the government but has also reduced the irrational fear that free speech creates in the minds of the people and the authorities. The Court's rulings have encompassed the people's right to criticize government officials, to picket in labor disputes, to protest and demonstrate against racial, gender, and sexual discrimination, to protect their privacy, to protect their reputations, to enjoy adult movies and television programs, and to use commercial speech for economic and political purposes. The American people did not know that they had so many rights until the Supreme Court began to think about the meaning of the First Amendment, and the thinking is not over as we enter the postmodern digital age.

There is no aspect of American society that has not been touched by the Court's interpretation of the First Amendment in relation to other rights. Fortunately, the Court has been interpreting the meaning of freedom of speech without subscribing to any particular theory of the First Amendment; consequently, the Court's openness has enabled it to deal with many complex situations in creative ways, which would not have been possible had it put itself into a theoretical or philosophical straitjacket. This does not mean that the Court's decisions have been ad hoc; rather, the Court has been basing its thinking on a set of assumptions that question what unfettered speech can do for the individual and society. Over time, its decisions have maximized the freedom of speech and other values. Depending on the situation, the Court has used in its free speech rulings the concept of the bad tendency, clear and present danger, the balancing of competing interests, the preferred position, and the affirmative and positive value of the First Amendment. Based on these assumptions, the Court has ruled that the First Amendment does not guarantee absolute protection to speech, which is consistent with the Fifth and Fourteenth amendments, under which liberties can be constrained, but only after due process.

UNFETTERED SPEECH

The Supreme Court's free speech decisions have taken into account the generative possibilities of unfettered speech, as well as what harm the government's restraints against speech might do. Throughout the ages, authorities' concern about free speech has been based on the fear that words have the power to instigate undesirable thoughts in people and to undermine their trust in the government, which might lead to rebellion and social disorder or corrupt the morals of impressionable and vulnerable members of the society. If free expression engenders a bad tendency, it is best to curb it before it does any harm. The bad tendency theory was the basis of the Sedition Act of 1798, which was used to convict mostly Republican editors. It was also used to convict Socialists and Communists during World War I and forms the basis of current obscenity laws. The Court defined obscene expression as that which arouses lascivious thoughts or appeals to one's "prurient interests" and is devoid of literary, artistic, political, or scientific value, as we will discuss later.

But the bad tendency test of free speech has not been used widely for censoring expression in the United States since Justice Oliver Wendell Holmes created a new legal standard, "the clear and present danger test," in the landmark 1919 case *Schenck v. United States* and gave the test an operational definition of "shouting fire" in a crowded theater. Charles Schenck, the general secretary of the Socialist Party, and another party worker printed and distributed antiwar leaflets urging people not to respond to the draft during World War I. They were found guilty of violating the Espionage Act and sentenced to prison terms, but they appealed to the U.S. Supreme Court complaining that their First Amendment right of free speech had been violated. By the time the

Court took up the case, the war had ended in victory for the United States and its allies. Justice Holmes, writing for the unanimous court, stated that the leaflets presented "a clear and present danger" to the government's recruitment efforts and hindered the nation in its efforts and preparation to win the war. The clear and present danger test was an improvement over the bad tendency test because it did not involve prior restraint on free expression. Justice Holmes summed up his central idea as follows: "The most stringent protection of free speech would not protect a man in falsely shouting fire in a theater and causing panic."[35]

A deliberate falsification of facts, if it led to panic and endangered lives, would be an example of a clear and present danger, but if the speaker actually saw fire and the warning saved lives, it would be protected speech. Moreover, if Schenck had simply expressed an opinion about the war without urging resistance to the draft efforts, his speech would have been protected under the guarantees of the First Amendment. But through his leaflets, he was preventing the government from fulfilling its constitutional obligation of raising an army to fight the war. In other words, when expression becomes an action or leads to an action that prevents the government from protecting the vital interests of the nation, the speaker can be punished. Thomas I. Emerson discusses the expression-action boundary in *The System of Freedom of Expression:*

> The central idea of a system of freedom of expression is that a fundamental distinction must be drawn between conduct, which consists of "expression," and conduct, which consists of "action." "Expression" must be freely allowed and encouraged. "Action" can be controlled, subject to other constitutional requirements, but not by controlling expression. . . . The definition of "expression" involves formulating in detail the distinction between "expression" and "action." The line in many cases is not clear. But at some points it becomes obscure. . . . In these cases it is necessary to decide, however artificial the distinction may appear to be, whether the conduct is to be classified as one or the other.[36]

Balancing competing interests is another approach that the Court has adopted in deciding whether free speech deserves protection. Zechariah Chafee Jr. writes in *Free Speech in the United States* that freedom of speech is one of the constitutional interests that has to be protected, but when it comes into conflict with other fundamental rights, it becomes

necessary to evaluate which interest is the most important in a given circumstance.[37] For example, the Court held in *American Communications
Association v. Douds* that a labor union official's right under the First
Amendment to refuse to disclose under oath whether he was a member
of the Communist Party, in an era when the Soviet Union was on the
rise and relentlessly preaching, aiding, and abetting worldwide bloody
revolution, did not outweigh the government's interest in preventing
strikes in the defense industries.[38] The ruling would be different today
because communism is no longer a threat, but the same perspective
could be brought to bear on a situation that threatened another vital public interest; for example, a state legislature could enact a law that prevented political campaigning near polling stations, provided the law
was not content based and advanced a compelling interest of the government.[39]

Despite the acknowledgment that government might have a compelling interest in restricting expression in specific circumstances, the
Court has accepted the view that freedom of speech must have a preferred
position in a democratic society. According to the preferred position theory, any restriction on free speech is unconstitutional unless proven otherwise; therefore, the government carries a heavy burden in justifying its
actions in restricting speech. Normally, in constitutional cases the burden
lies on the individual to prove that his constitutional rights have been violated; in free speech cases, it is the other way around—the government
must prove that restraint on expression protects an overriding interest of
the government. The government works under a "heavy presumption"
that restraining speech is unconstitutional and therefore "bears a heavy
burden of showing justification for such restraint."[40]

The Court has developed a three-step approach in deciding free
speech cases: (1) the government must show a compelling constitutional interest in restraining speech, (2) the law must not be overbroad
and must be specific in its focus so as to cause the least harm to free
speech, and (3) the government cannot protect the vital interest in question except by restraining free speech.[41] The preferred position brings
free speech back almost to the literal wording of the First Amendment,
"Congress shall make no law," but as we have discussed, other interests
and obligations must be taken into account.

The Court's interpretation of the First Amendment has given as much
protection to peaceful demonstration, including symbolic expression, as

to political expression. In several cases, the Court has ruled that when acts are committed to express an opinion on a public issue, they are protected symbolic speech. To receive First Amendment protection, symbolic expression must have a message for the audience, even if the symbolic act is socially disapproved of and considered abominable by some, for example, the burning of the national flag.[42] Symbolic speech does not include criminal acts, such as murder, assault, and battery, with a view to communicating a message, even if the message is an implicit one. When a person burned a draft card, the act was not considered symbolic speech because it prevented the government from fulfilling its constitutional obligation of protecting the country by recruiting for the armed forces. But when someone carried a placard saying, "F*** the draft," that was deemed protected symbolic speech.[43] Conduct as symbolic speech receives First Amendment protection if the message and the conduct are so integrated that they cannot be separated and relate to a public issue. But what if the speech-conduct symbolic expression is posted anonymously on the Internet?

LIBEL AND FREE SPEECH

The Internet has begun to test the legal and ethical boundaries of free speech. Many people believe that they can take advantage of the anonymity provided by online service providers, post any message, and get away with it because no one will be able to trace them.

The laws of terra firma do apply to cyberspace, though it may not be always possible to catch violators if they encrypt their messages and use a chain of anonymous remailers. The opacity of cyberspace is an illusion, which might have been fostered by Peter Steiner's July 1993 *New Yorker* cartoon showing a dog chatting with another dog via computer and saying, "On the Internet, nobody knows you're a dog." But if anyone listens to a dog, naturally they will end up in a doghouse.

Dr. Sam D. Graham, a former professor and chair of the urology department at the Emory University School of Medicine in Atlanta, Georgia, resigned from his position in July 1998, moved to Richmond, and started a private practice. An anonymous message on Yahoo!'s message board, which Dr. Graham saw in February 1999, accused him of receiving kickbacks from a urology company that was doing business with his former department at the medical school. A doctor has nothing but his reputation to live on. When his attorneys filed a libel suit, Yahoo! was forced to disclose, under the discovery process, the name of the anonymous author, who turned out to be another pathologist, Dr. Jonathan R. Oppenheimer of Prost-Data. A U.S. district court judge awarded Dr. Graham $675,000 in damages for defamation. The anonymous Internet message, the judge said, was "about as despicable as any course of conduct that one could engage in."[44]

Consider the case of AK Steel of Middletown, Ohio, a publicly traded company, which became the target of defamatory pseudonymous messages in 2000 posted on the Yahoo! message board by some Jane Doe. The company's counsel, John Hritz, who was also a target of the defamatory messages, wanted Yahoo! to disclose the identity of Jane Doe so that he could proceed successfully with a libel case, but civil liberties groups, including the Electronic Frontier Foundation and the Public Citizen, defended the anonymity of Jane Doe, criticizing the company for trying to silence its critics.[45]

Anonymous and pseudonymous communication enhances freedom of speech and is protected by the First Amendment, but when this right conflicts with the Sixth Amendment right to a fair trial, the courts have to strike a balance between the two. The courts can force a service provider to disclose a person's identity so that there is no miscarriage of justice, though there is also a danger that subpoena power could be used to silence anonymous or pseudonymous criticism. Since AK Steel is a publicly traded company, under the law the company has the status of a public person and therefore must meet a higher burden of proof, the "actual malice standard," in order to win damages. "Actual malice" was defined by the U.S. Supreme Court in a landmark case, *New York Times Co. v. Sullivan* (1964), as reckless disregard for the truth or knowingly publishing misleading information, regardless of motive. There was no way for AK Steel to meet the burden of proof unless its attorneys were allowed to question the accuser, Jane Doe, who protected her identity under the right to anonymous speech in the public interest. Sometimes a company finds that no response is the best response to anonymous defamatory speech, though the threat of legal action against ISPs makes them write clear policies for the use of their services.

The Internet anonymity provided by ISPs gives their service users a false sense of security. It is an obligation of online service providers to warn the users that in cyberspace they are as "naked as in a fishbowl."[46] The only way a person can be assured of strong security regarding the anonymity of his messages is to use strong cryptography and then send the message through a chain of remailers (hidden like a Chinese box within a box) located somewhere outside the country. Remailers send messages from nobody@nowhere, and those based outside the United States are under no obligation to obey U.S. law or to surrender information about or the identity of the sender.

Every society recognizes the value of reputation and how it can be damaged by words that accuse a person of crime, immorality, or professional dishonesty. For a long time in Britain and the United States, courts under the common law allowed the victims of oral (slanderous) and written (libelous) defamation to be paid monetary compensation for the loss of reputation. But as mass media began to create an environment of fast and free-flowing information, it became impossible to localize the damage to one's reputation. People began to realize with horror that the medium, as a system of networked communications, has the potential to spread defamatory statements far beyond one's circle of friends and acquaintances. Libel laws seemed to provide protection against the excesses of the press and were extensively used to seek redress until the cost of litigation threatened to silence the press, and the Supreme Court once again stepped in to rescue freedom of speech and expression. Although the Court's libel decisions did not end libel suits, they raised the legal standards for winning libel damages.

In *New York Times v. Sullivan*, the Court ruled that the First Amendment protects the press from libel suits brought by public officials, even if the statements published are false, unless the official can prove that the statement was published in reckless disregard for the truth. The burden of proof shifted from the press to the plaintiff, the public official who must prove that the press wrote a reckless lie. With this and subsequent libel decisions that also included public figures and celebrities, the Court raised libel from the realm of common law to the higher ground of constitutional law and set the standard of proof high enough to save the media from self-censorship. Rampant libel suits could have killed freedom of the press, reducing the media to mere conduits for rehashed press releases by government and corporations.

The common law of defamation, which was based on tradition and communal practices in England, was brought to the American colonies by colonial rulers and English settlers and was adopted by various states after the American Revolution. Until 1964, the common law of defamation, based on seven hundred years of history in England and the United States, favored those who claimed that they had been defamed. Before *New York Times v. Sullivan*, the media carried the burden of proof and could protect itself from damages by proving that the defamatory statement was based on truth or a fair comments privilege. Even a claim to truth was an insufficient defense in some cases if the defamed person could prove ill will on the part of the media. Fair comment had to be based on facts, and the privilege was limited to actual reporting of government or judicial facts. In this sense, the *New York Times* case was transformative because it overturned this tradition, thereby immensely extending the domain of free speech in American society. The Court ruled that, regardless of the motives of the speaker, truth can never be libelous. Opinions and comments on matters of public concern are constitutionally protected unless they are factual assertions that can be proved false or they threaten imminent lawlessness. Hyperbole, satire, put-downs, opinions, comments, and the like are protected. The Court's ruling reduced libel to a false and defamatory assertion of facts, told with full knowledge that they were false, which caused measurable harm to the plaintiff. The Court reasoned that the media could go beyond press releases and handouts to dig deeper and uncover truth about the public performance of officials only if the debate was "uninhibited, robust and wide open."[47]

The genesis of the *New York Times* case lay in the struggle of the blacks against segregation, especially in the South, where they had separate bus seats, lunch counters, water fountains, and other public facilities. In 1960, inspired by the struggle and leadership of Martin Luther King Jr., black students at Alabama State College in Montgomery held a demonstration to end segregation in the school. Supporters of the civil rights movement published a full-page advertisement in the *New York Times*, which stated that the state police had committed atrocities against students, including padlocking the student dining hall "in an attempt to starve [the students] into submission." The editorial advertisement also accused the police of bombing the home of Dr. King, "almost killing his wife and child," assaulting him, arresting him seven times for offenses such as speeding and loitering, and charging him with perjury.

Since the reference in the advertisement to the state police really meant the police commissioner, Police Commissioner L. B. Sullivan filed a suit under the Alabama libel law. The advertisement was neither an accurate statement of facts nor a fair comment. Nor was it protected by the privilege of reporting already known facts from public records. According to the Alabama libel statute, the case was clear and convincing beyond any doubt that the *New York Times* had committed libel against L. B. Sullivan, and the state supreme court affirmed damages amounting to $500,000 against the newspaper. This was not the only libel suit against the media. CBS, for example, was a defendant in another case that involved reporting on racial segregation. Libel was being used as a tool to circumvent the First Amendment, and under these compelling circumstances, the U.S. Supreme court gave a momentous decision that pushed the meaning of freedom of speech and expression to new horizons.

Justice William Brennan, in a 5–4 decision, wrote that a political advertisement, unlike commercial speech selling a product, is protected under the First Amendment. Freedom of expression should not be limited because the expression is published "in the form of a paid advertisement." Even if the statement is libelous per se, it must be examined in the light of First Amendment freedoms. Justice Brennan wrote,

> Thus we consider this case against the background of a profound national commitment to the principle that debate on public issues should be uninhibited, robust, wide open, and that it may well include vehement, caustic, and sometimes unpleasantly sharp attacks on government and public officials. The present advertisement, as an expression of grievance and protest on one of the major public issues of our time, would seem clearly to qualify for the constitutional protection.[48]

Since the advertisement contained a false assertion of facts, the question was how far the First Amendment could go to protect such assertions. The Court said that in the heat of the debate on public issues, it was likely that some falsehood might be uttered, and "it must be protected if the freedoms of expression are to have the 'breathing space' they need . . . to survive."

The only way, Justice Brennan wrote, a public official could expect to recover damages from a libel suit was if it was proved that "the statement was made with 'actual malice,' that is, with knowledge that it was false or with reckless disregard of whether it was false or not." The expression "actual malice" has caused lot of confusion. It does not mean evil intent but knowledge on the part of the media that a statement is false or might be false, and yet they published it anyway. The Court shifted the burden of proof to defendants, the public officials or public figures who claim that their reputations have been damaged, and they must prove this damage with "convincing clarity," beyond any reasonable doubt, which is the standard applied in murder cases. Negligence, or failing to take reasonable care, which was the *New York Times* advertising department's culpability as the statement should have been checked, is not actual malice.

In a subsequent libel case, *Gertz v. Robert Welch, Inc.*, the Court extended the concept of public officials to include public figures, those people who thrust themselves "into the 'vortex' of an important public controversy." The Court also clarified further that defamatory statements are actionable only if they make false assertion of facts. Writing for a 5–4 majority, Justice Lewis F. Powell said,

> Under the First Amendment there is no such thing as a false idea. However pernicious an opinion may seem, we depend for its correction not on the conscience of judges and juries, but on the competition of other ideas. But there is no constitutional value in false statements of facts. Neither the intentional lie nor the careless error materially advances society's interest in "uninhibited, robust, and wide-open debate" on public issues. . . . They belong to that category of utterances which are no essential part of any exposition of ideas, and are of such slight social value as a step to truth that any benefit that may be derived from them is clearly outweighed by the social interest in order and morality.[49]

The *New York Times* case created a superior category of speech dedicated to the discussion of public affairs, which was given near absolute protection and is increasingly being used in the postmodern digital age to make public, as well as corporate, affairs transparent. But for business communicators, public relations practitioners, advertisers, employers, credit bureaus, and other nonmedia organizations, the common law is still applicable, and the burden of proof still lies with the defendant, unless the nonmedia communication concerns a matter of public interest.

OBSCENITY AND THE PARADOX OF FREE SPEECH

The courts in the United States have been spending enormous amounts of judicial time and intellectual resources in grappling with the problem of sexual explicitness in order to discover genuine human expression buried in obscenity that deserves to be protected by the First Amendment. Concerns about sexual expression have been essentially a question of protecting the morals of the people, especially the most vulnerable segment of society, children.

The Supreme Court has struggled to determine the quality and level of expression that is necessary to lift obscene material to the standard of constitutionally protected speech. But the problem of distinguishing sexual expression as a protected speech from nonprotected sexual conduct has become more complicated because now it is mostly the marketplace that is driving all forms of sexual manifestation in art, literature, movies, and advertising. This puts the Court in a difficult "position much like that of a child trying to throw away an unwanted boomerang."[50]

When in 1997 the U.S. Supreme Court struck down sections of the Communications Decency Act (CDA), which would have criminalized making indecent material available to children on the Internet, many people felt that the Court had embraced the Internet as a unique medium of communication. Not fully aware of the full potential of the new interactive medium, which is evolving at a baffling speed and engulfing all fields of human endeavor, the Court accepted the fallacious argument that the Internet is more like newspapers, books, or pamphlets; therefore, any attempt to tamper with it by the government requires strict scrutiny. The Internet deserves the highest level of protection accorded by First Amendment freedoms.

The Internet has a deceptive resemblance to the older mass media, but in reality its decentralized freedoms, chaotic boundlessness, and freewheeling exchange make it altogether a different mode of communication. The Supreme Court, at a loss for an analogy to illuminate our understanding, nonetheless amused us with rich-sounding words, reminiscent of the nation's early history when every American was his own pamphleteer and a soapbox orator. "Through the use of chat rooms, any person with a phone line can become a town crier," said Justice John Paul Stevens for the Court majority. "Through the use of Web pages, mail exploders and newsgroups, the same individual can be a pamphleteer." That's a dangerous half-truth, because Internet surfers are not a political crowd or an assembly; rather, they are lonely, dispersed aggregates with little sense of the community, except in the digital sense. You cannot be a pamphleteer without a community, though it is possible to arouse awareness and activate a community if it already exists or has a potential existence, as can be seen in the case of some bloggers or websites like MoveOn.org and Second Life.

There were some practical difficulties with the CDA. It lacked "the precision that the First Amendment requires when a statute regulates the content of speech." The Supreme Court has always been chary of giving the government broad powers to restrict speech, unless there is a clear and present danger, such as an imminent threat to the peace or to national security. The government does have a legitimate "interest in protecting children from harmful materials," said Justice Stevens, but he admonished, "In order to deny minors access to potentially harmful speech, the CDA effectively suppresses a large amount of speech that adults have a constitutional right to receive and address to one another."[51]

In several cases, including *FCC v. Pacifica Foundation* (instigated by the comedian George Carlin's monologue "Filthy Words," discussing seven words that "you couldn't say on the public airwaves"), the Supreme Court has held that the First Amendment protects offensive speech, including sexual "expression, which is indecent but not obscene."[52] The government does not have the authority to reduce adult speech to a level fit for children. The level of "discourse reaching a mailbox simply cannot be limited to that which would be suitable for a sandbox," said Justice Stevens.[53] But trying to prevent a child playing in a sandbox from opening an adult mailbox containing sexually explicit material has opened new dangers of censorship by private groups with their own moral and political agendas. With the touch of a key, anyone equipped with filtering technology can block any file from a website deemed objectionable. Filtering technology is not value neutral.

Consider how the marketplace and industry, in response to social and political pressures, are trying to make self-regulation work to control the flow of pornography and obscene materials. One such method of self-regulation is software like the Platform for Internet Content Selection (PICS). PICS is more than a technical filter because not only can it block access to pornographic and violent content but it can also be used to erase information about abortion and birth control or even a legitimate website by an educational group or commercial entertainer. Authoritarian regimes like Saudi Arabia, Singapore, and China have been using filtering software to prevent the free flow of information to their captive millions. A program like PICS could become a silent killer of information, a blocker of arteries through which information flows, a universal censor.[54]

Can the U.S. Supreme Court prevent the making of a technology that abridges the freedom of speech and expression? Now we are talking about the power of the judiciary vis-à-vis the marketplace, the global marketplace. By preventing the government from regulating sexual content on the Internet to protect children, the Supreme Court might have unwittingly passed the right to censor speech on to a few media giants or well-organized social, religious, and political groups, who could block out not only the obscene but also ideas that they find offensive. When the government censors speech, you can go to the court for remedy. But what can you do if a service provider or a religious organization, such as a church or a mosque, uses PICS or some other software to censor information? You would not even know that some information had been deleted or blocked.

The freedom to read is not stated explicitly in the Constitution, but it is deemed a concomitant part of freedom of expression, which, like other rights, is not absolute. While expression can be controlled through prior restraint, reading can be subjected to variable controls ranging from banning obscene books to limiting access by zoning laws and other means, such as digital rights management. The fear of creative expression and erotica has an ancient history, and only in civilizations where erotic expression served spiritual purposes, as in ancient India, for example, was it allowed to flourish. In the age of the Internet, expressive erotica serves the marketplace, but it is also protected by the First Amendment principle—first do no harm to free expression—which puts an enormous burden on the government to prove the necessity of doing so.

THE *MILLER* STANDARD

Although the words "pornography" and "obscenity" are sometimes used synonymously in discussing First Amendment protections for sexual expression, obscenity has been given a legal meaning by the U.S. Supreme Court as denoting material that is sexually explicit and is made for sexual arousal and amusement but that has no socially redeeming value. Indecency is a class of objectionable speech that is applicable to broadcasting and can be restrained under the provision of time, place, and manner, even though it may not be obscene.

Although obscenity was not defined rigorously, by the middle of the nineteenth century, many states had enacted statutes to prosecute obscenity cases. At the federal level, Anthony Comstock, a veteran of the Civil War from New York, was a most zealous crusader against obscenity and lobbied Congress to enact in 1873 the first federal postal law against mailing obscene material. In a modified form, the Comstock Act is still on the statute books.[55] Before the Supreme Court began to ponder the meaning of sexual expression in the late 1950s, obscenity prosecutions had led to the banning of some serious literary works, such as Sinclair Lewis's *Elmer Gantry*, Theodore Dreiser's *An American Tragedy*, D. H. Lawrence's *Lady Chatterley's Lover*, and James Joyce's *Ulysses*, among others.

In a 1957 case, *Roth v. United States,* the U.S. Supreme Court made the first attempt to grapple with obscenity and established guidelines to distinguish it from constitutionally protected expression.[56] With this foundational principle of obscenity law, the Court lifted literary works, art, movies, and other forms of human expression from the category of unprotected speech, unless the prosecution could show that it had no socially redeeming value. In distinguishing obscenity from an artistic portrayal of sex, the Court said,

> [S]ex and obscenity are not synonymous. Obscene material is material, which deals with sex in a manner appealing to prurient interest. The portrayal of sex, e.g. in art, literature and scientific works, is not itself sufficient reason to deny material the constitutional protection of freedom of speech and the press. Sex, a great and mysterious motive in human life, has indisputably been a subject of absorbing interest to mankind through the ages; it is one of the vital problems of human interest and concern.[57]

So long as sexual portrayals could be integrated with expressive ideas, the expression was protected by the First Amendment, a strategy that *Playboy* used to present itself as a magazine of serious ideas wrapped in the body of a lusty, desirable woman. Moreover, in the 1950s, sexual mores had begun to change, and Americans had become more tolerant of openness in sexual expression. The Court, as it normally does, reflected the mode of consciousness of the contemporary society. What in earlier times was banned as obscene, for example, John Cleland's *Memoirs of a Woman of Pleasure*, popularly known as *Fanny Hill* (published in 1820), became acceptable when the mode of consciousness changed in the late 1950s and 1960s, and it was discovered that the book had artistic merit, historical value, and implicit moral sensibility.[58] The Court ruled that the jury could not find a work obscene unless the prosecution proved that the work had no uplifting social value. This overbroad approach to obscenity due to changing views of sexual expression led to several ad hoc case-by-case decisions until in 1973 the Supreme Court reshaped the law in *Miller v. United States* and established standards that rule today.

Marvin Miller was convicted in 1973 of violating California law by mailing unsolicited, sexually explicit material to a restaurant owner who reported the matter to the police. When the case finally reached the Supreme Court, it presented an opportunity for the Court to rewrite and refine the earlier definition of obscenity and to enunciate clear standards for obtaining convictions by establishing a three-part test. Justice J. Warren Burger, writing for the majority, established the following three-part criterion for deciding an obscenity case:

1. "The average person, applying contemporary community standards," will "find the work, taken as a whole, appeals to the prurient interest."
2. "The work depicts or describes, in a patently offensive way, sexual conduct specifically defined by the applicable state law."
3. "The work, taken as whole, lacks serious literary, artistic, political, or scientific value."[59]

Patently offensive portrayal of sex, according to the Court, included, for example, "descriptions of ultimate sexual acts, normal or perverted, actual or simulated . . . masturbation, excretory functions, and lewd exhibition of the genitals." Representations of sexual expression "must have serious literary, artistic, political, or scientific value to merit First Amendment protection." Acknowledging that the standard would vary from community to community, the Court rejected the idea of one national standard for obscenity and left it to local juries to determine, in keeping with state law and the standard of decency, what material was obscene. As Justice Burger wrote,

> It is neither realistic nor constitutionally sound to read the First Amendment as requiring that the people of Maine or Mississippi accept public

depiction of conduct found tolerable in Las Vegas, or New York City. . . . People in different states vary in their taste and attitudes, and this diversity is not to be strangled by the absolutism of imposed uniformity.[60]

Although trust in local community standards has been the foundation of American political culture, the due process provisions of the Fourteenth Amendment and other constitutional prohibitions mitigate the dangers of the excesses of localism and isolationism. Justice William O. Douglas's fear, expressed in his dissenting opinion, that the ruling "would make it possible to ban any paper or any journal or magazine in some benighted place" has been unfounded.

State prosecutors have used various legal alternatives to criminal obscenity statutes, such as zoning laws, nuisance laws, public indecency laws, and racketeering laws in order to control obscene conduct and expression.[61] But such attempts have been constrained by the Court under the guidelines developed in *United States v. O'Brien*, which include the following: the government must have an important or substantial interest in preventing the activity, the government interest must not be directly aimed at the suppression of free expression, and the restriction of First Amendment freedoms must not be more than is absolutely essential to furthering the government's specific interest.[62]

In short, police and prosecutors cannot act arbitrarily and seize the material containing sexual portrayals without first going to the court and proving it legally obscene. Under these rules, the performance of the 1968 love-rock musical *Hair* could not be stopped in a municipal auditorium, despite the fact that it contained simulated sexual acts, as well as nudity.[63] When Broward County, Florida, sheriff Nicholas Navarro got a restraining order from a local judge to prevent the sale of 2 Live Crew's album *As Nasty as They Wanna Be*, the appeals court ruled that intimidating music retailers with a probable cause notice was tantamount to prior restraint.[64]

For about a half-century, the Court has struggled with the problem of obscene expression; nonetheless, it has not reached a satisfactory conclusion as to whether the "delivery of obscene messages to consenting adults may be prosecuted as a crime" or "criminal prosecution for obscenity-related offenses violates the First Amendment."[65] The Supreme Court cannot rise above the divisions in society. All the Court does is end the dispute by its ruling, until the social pressure rises enough for the Court to revisit the problem.

Protecting Children from Indecency and Pornography

It is difficult to comprehend why some adults see children as sex objects. Traffic in child pornography led Congress to enact in 1977 the Protection of Children against Sexual Exploitation Act, which provides for the prosecution of people who make or sell photographs of children engaged in sexual acts.[66] All fifty states have child pornography laws that delineate the crime of child pornography in specific sexual acts. Even an isolated portrayal of a sexual act involving a minor can be prosecuted under the act, regardless of the provisions of the *Miller* standard, which tolerates obscenity if the work has some redeeming social value. In *New York v. Ferber*, the Supreme Court upheld the state law against child pornography and concluded that "the use of children as subjects of pornographic materials is harmful to the physiological, emotional, and mental health of the child. The judgment, we think, easily passes muster under the First Amendment."[67]

However, at least two justices considered it possible that a work portraying sexual activity with teenagers might be deemed art and might need constitutional protection under certain circumstances.[68] It must be noted that no federal or state laws have been enacted to prevent minors from having sex; rather, legislation about children's sexuality has aimed at the commercial exploitation or expression in any form of children's sexual activities. The possession of child pornography even in the privacy of one's home does not give a person protection from prosecution under the law, though a person may read or watch pornographic material involving adults in the privacy of his home.[69]

The U.S. Supreme Court has established different standards regarding indecency and obscenity for various communications media. Because of the unique nature of broadcasting, for example, its ability to reach unwilling viewers, including children, the Court has frowned upon "indecent" programming for broadcasting; consequently, its definition of "indecency" is much narrower than that prescribed for obscenity in *Miller v. California*. Under the narrower standard, broadcasters can be penalized for airing programs with sexually explicit language and portrayals, except during a time period called a "safe harbor," a restriction that is not applicable to print, cable, theater, movies, nightclubs, and the Internet. Broadcasters thus have less freedom of speech than other media because of broadcasting's easy accessibility to children, but these restrictions might be construed to fall under the rubric of time, place, and manner restrictions.

As mentioned earlier, a landmark ruling that upheld the Federal Communications Commission's (FCC) right to punish indecent expression in the broadcast media arose from a case in which a disc jockey at the Pacifica Foundation's FM radio station in New York played, at a time when kids are home from school, George Carlin's "Seven Dirty Words" routine, which the stand-up comedian presented before a nightclub audience.[70] The constitutional question before the Court was whether the FCC had the right to regulate the station's broadcasting of "filthy words" under 18 U.S.C. 1464 (which prohibits the airing of "obscene, indecent, or profane language") without due process and without violating the station's First Amendment rights. In essence, the Court had to rule whether what was good for a mature adult audience was equally good for a school audience. The Court ruled that while the "Seven Dirty Words" monologue was not obscene, words could be indecent or profane without being obscene. The Court's plurality defined "indecent" as "nonconformance with accepted standards of morality" and upheld the FCC's decision against the radio station. Although the Communications Act prohibits the FCC from censoring expression, even if it is obscene, indecent, or profane, the Court ruled that the commission could punish a station for broadcasting such material because its accessibility threshold is low (for children) and therefore it cannot be treated on a par with other media.

Justice John Paul Stevens, writing for the majority, gave two reasons for upholding the FCC's decision to punish indecent programming. First, broadcasting is an intrusive medium because it enters the privacy of a person's home unannounced, and home is "where the individual's right to be left alone plainly outweighs the First Amendment rights of an intruder . . . [where] prior warning cannot completely protect the listener or viewer from unexpected program content. To say that one may avoid further offense by turning off the radio when he hears indecent language is like saying that the remedy for an assault is to run away after the first blow." Second, Justice Stevens said, "broadcasting is uniquely accessible to children, even those too young to read."[71]

Parents have an obligation to protect their children from indecent exposure from broadcasting media, and they expect the government to help them.[72] But when Congress passed a statute in 1988 ordering the FCC to enforce the antidecency ruling on broadcasting twenty-four hours per day, a serious constitutional question was raised. Such a blanket ban on indecent expression could be valid only if it served a compelling public interest in safeguarding children. In 1991, the Federal Court of Appeals ruled on the question of a twenty-four-hour ban on indecent programming in *Action for Children's Television v. Federal Communications Commission,* reinterpreting *Pacifica* broadly to include a provision for allowing adult programming during a safe-harbor period to be determined by the FCC (10 p.m. through 6 a.m.).[73]

Broadcasters interpreted the Court's ruling in *Pacifica* as permitting them to broadcast raunchy materials on public airwaves at times when children were not listening; in short the ruling did not authorize the FCC to control content except indirectly through time, place, and manner restrictions. The FCC rewrote its definition of indecent expression to include "language or material that, in context, depicts or describes, in terms patently offensive, as measured by contemporary community standards for the broadcast medium, sexual or excretory activities or organs." This definition shifts the emphasis from the whole community to the broadcast medium community and from the whole of the message to a part of the message, for example, a single obscene word or a gesture. Also, there is no reference to the broadcast's appealing to a prurient interest in sex. This is a modified version of the *Miller* test for obscenity.

For constitutional reasons, the courts have not extended the *Pacifica* ban regarding indecent expression to cable or telephone communications, because the user has better control over the media.[74] Consequently, cable viewers may have greater exposure to nudity, indecent expression, and sexually explicit language than broadcast television or radio audiences. Similarly, the Supreme Court ruled in *Sable Communications of California v. FCC* that indecent expression that is not obscene is protected by the First Amendment and may be curtailed only if the ban is tailored narrowly to serve a compelling government interest.[75] The government cannot deprive adults access to programs unsuitable for children, such as Dial-a-Porn, by instituting a total ban when less offensive solutions are available.

Sexual Expression and the Internet

Reports about the widespread availability of sexually explicit material in cyberspace, a medium that has been increasingly popular with children, almost created mass panic and led to constitutionally weak and hasty legislation in the later part of the 1990s.[76] Many people felt outraged that sex-oriented chat rooms and pornographic websites were invading their homes and that their children might access those sites serendipitously. But those who saw the untapped potential of the Internet, the nascent, global medium of spontaneous mass communication that was pushing the boundaries of human freedom, saw a great threat in the government's preemptive action. As noted earlier, through its rulings in several landmark cases, the Supreme Court has accorded expression in various categories of speech and media different levels of protection under the First Amendment. The print media, newspapers, magazines, and books, for example, have the greatest freedom. On the other hand, because the number of radio frequencies available is limited, and because it uses a scarce public resource and is intrusive and easily accessible to children, broadcasting is subject to content regulation through time, place, and manner restrictions and other measures in the public interest.[77] The Supreme Court has accorded cable television a status akin to that of newspapers, and it has greater First Amendment protection than broadcasting.[78]

Under the *Miller* ruling, obscenity is an unprotected form of expression under the First Amendment, and it may be banned on the Internet and other media if the legislation is narrowly tailored to serve an overriding government interest. But the problem is that most people, including lawmakers, dump pornography, obscenity, and indecency into one and the same category of speech. While pornography is a subset of obscenity, everything that is legally obscene may not be pornographic, yet it may be banned under the *Miller* rule. Indecency, as a category of nonconformative speech, is a protected expression in the print media and cable, but under the *Pacifica* doctrine, it is not protected speech in broadcasting, where it may be regulated. The First Amendment status of the Internet as a medium of communication was tested for the first time in *Reno v. ACLU* in 1997.

The case arose out of the Communications Decency Act (CDA), a portion of the Telecommunications Act of 1996, which was signed into law to ban from cyberspace any sexual expression that was patently offensive or indecent, though not necessarily obscene, unless the viewing of the indecent expression could be blocked from minors.[79] As we noted earlier, the Court upheld a partial federal ban on indecent expression in broadcasting under the *Pacifica* ruling (through the safe-harbor provision) because the medium was invasive and could catch children unintentionally. In *Reno v. ACLU,* the Court ruled that the Internet was not as intrusive as broadcasting because the user has to take deliberate steps and search for websites, the way they have to access Dial-a-Porn on the telephone. Justice John Paul Stevens, writing for the Court, said that the CDA was vague and overbroad because "indecency" and "patently offensive" expression as measured by contemporary community standards for the discussion of sexual or excretory activities were not defined. The lack of definition would create "uncertainty among speakers about how the two standards [indecency and patent offensiveness] relate to each other. . . . This uncertainly undermines the likelihood that CDA has been carefully tailored to the . . . goal of protecting minors."[80] To pass constitutional muster, the law should have been written precisely to avoid being overbroad and vague. In keeping with constitutional tradition, Justice Stevens concluded, "we presume that governmental regulation of the content of speech is more likely to interfere with the free exchange of ideas than to encourage it. The interest in encouraging freedom of expression in a democratic society outweighs any theoretical but unproven benefit of censorship."[81]

Congress's response to the Supreme Court ruling was the adoption in 1998 of the Child Online Protection Act (COPA), which differed from the CDA in three respects. First, the COPA banned only speech that is "harmful to minors" in the sense of *Miller*'s obscenity standard. The CDA, on the other hand, banned all indecent speech that was "available to minors." Second, the COPA restriction was applicable to speech used "for commercial purposes," while the CDA applied to any kind of speech. Third, the COPA allowed defendants to use an "affirmative defense"; for example, they could prove that they required users to show an ID for age verification, such as a credit card or a driver's license. The CDA did not have this provision.

A free speech coalition led by the American Civil Liberties Union challenged the COPA and got a preliminary restraining order from U.S. District Court Judge Lowell Reed Jr., who ruled that the use of credit card or age verification methods for accessing "material that is harmful to minors" might hold people back from accessing such material and might have an adverse economic effect upon service providers. Reference to "adverse economic effect" means that the marketplace was an important deciding factor in the free speech case.[82] The Third Circuit Court of Appeals unanimously upheld the lower court's COPA decision, adding that the law "imposes a burden on speech that is protected for adults." The appeals court also saw a problem with "contemporary community standards" to identify content deemed harmful to minors because "any communication available to a nationwide audience will be judged by the standards of the community most likely to be offended by the message . . . [and] thus imposing an overreaching burden and restriction on constitutionally protected speech."[83]

Both the courts expressed the hope that filtering technology would offer an adequate means of protecting children from harmful speech without imposing an unnecessary burden on constitutionally protected speech for adults, but here also government's efforts have not met with success. In the Children's Internet Protection Act (CIPA), Congress in 2000 made another attempt to regulate Internet communications to protect children by making the availability of federal funding for libraries contingent upon their use of commercial filtering technology on all Internet terminals, but the American Library Association and others challenged the constitutional validity of the law. A special three-judge panel of the U.S. District Court for the Eastern District of Pennsylvania agreed that the government has a legitimate interest in "preventing the dissemination of obscenity, child pornography, or in the case of minors, materials harmful to minors, and in protecting library patrons from being unwillingly exposed to offensive, sexually explicit material, and for the First Amendment purposes, a public library's use of Internet filters." But the government, the court continued, also carries a heavy burden that requires it to tailor restrictions on free speech to serve a compelling state interest in a manner that is least offensive to the First Amendment. After considering various commercial filtering programs, the panel noted that at present no "technology can make the judgments necessary to determine whether a visual depiction fits the legal definition of obscenity, child pornography, or harmful to minors. Given the state of the art in filtering and image recognition technology, and rapidly changing and expanding nature of the Web, we find that filtering products' shortcomings will not be solved through a technical solution in the foreseeable future." The Court concluded that the use of CIPA-mandated filtering technology by public libraries is not narrowly tailored to further the government's interest in the issue and is therefore unconstitutional.[84]

Although the court ruled that the government may not ban or impose restrictions on Internet communication, the ruling in another case,

United States v. Thomas, suggests that the government can prosecute a person on obscenity charges in communities opposed to sexually explicit expression in certain circumstances. The case concerned a California couple, Robert and Carleen Thomas, who in 1994 started an Internet bulletin board service, the Amateur Action Computer Bulletin Board, which offered e-mail, chat lines, public messages, and sexually explicit videos and pornography to enrolled and paying subscribers. A federal postal inspector in Memphis, Tennessee, enrolled under a pseudonym and, after paying the membership fee, downloaded pornographic material. The couple was prosecuted under a federal law that banned the interstate transportation of obscene material. On appeal, the Thomases argued that since their business was located in California, the case should have been tried in California, where the San Jose Police Department had declared their erotica material not obscene under the state standard. Besides, they argued, the local community standard was not applicable to cyberspace, which has no boundaries. The U.S. Court of Appeals for the Sixth Circuit ruled in 1996 that the trial could be held only in Tennessee where the crime occurred. The Court rejected the idea of a broader cyberspace community standard, concluding that the bulletin board service could have denied service to districts where obscenity standards were different if the defendants had wanted to avoid prosecution.[85]

The U.S. Supreme Court voted 8–1 to reverse the appeals court's decision and remanded the case for further consideration in the light of its limited scope. Justice Clarence Thomas wrote for the majority that, "We hold only that COPA's reliance on community standards to identify 'material that is harmful to minors' does not by itself render the statute substantially overbroad for the purpose of the First Amendment. . . . We do not express any view as to whether COPA suffers from substantial overbreadth for other reasons, whether the statute is unconstitutionally vague, or whether the District Court correctly concluded that the statute likely will not survive strict scrutiny analysis once adjudication of the case is completed below."

Amidst this indecisiveness about COPA, Justice Anthony Kennedy, joined by Justices David Souter and Ruth Bader Ginsburg, expressed serious doubts as to whether COPA was consistent with the First Amendment. Their constitutional reservations became a majority opinion in a related case regarding virtual child pornography. The 1996 Child Pornography Prevention Act (CPPA) made it a federal crime to

include any visual description that "is, or appears to be, of a minor engaging in sexually explicit conduct" or that is "advertised, promoted,
presented, described, or distributed in such a manner that conveys the
impression that the material is or contains a visual depiction of a minor
engaging in sexually explicit conduct." The statute also prohibited the
use of youthful-looking adults posing as minors in sexual situations, as
well as computer-generated images of children engaging in sexual conduct without any actual child taking part. The Free Speech Coalition
challenged the CPPA on the grounds of overbreadth and vagueness.[86]

The U.S. Supreme Court agreed with the plaintiff that the act was
overbroad, and in a 6–3 decision struck down the CPPA as unconstitutional. Writing for the majority, Justice Kennedy said that the statute
"prohibits speech that records no crime and creates no victim by its production. Virtual child pornography is not 'intrinsically related' to the
sexual abuse of children, as were the materials in *Ferber*. While the
Government asserts that the images can lead to actual instances of child
abuse . . . the causal link is contingent and indirect." The law was not
limited to obscenity that is legally unprotected expression but applies to
"any visual depiction" that "is or appears to be of a minor engaging in
sexually explicit acts."

The statute, Justice Kennedy said, had the potential to ban speech that
might have redeeming social value, be it scientific, artistic, literary, or
political, as defined under *Miller*. Elaborating on the dangers of CPPA's
chilling free expression regarding teenage sexuality and the sexual
abuse of children, Justice Kennedy wrote that both themes "have inspired countless literary works":

> William Shakespeare created the most famous pair of teenage lovers, one
> of whom is just 13 years of age. In the drama, Shakespeare portrays the
> relationship as something splendid and innocent, but not juvenile. The
> work has inspired no less than 40 motion pictures, some of which suggest
> that the teenagers consummated their relationship. Shakespeare may not
> have written sexually explicit scenes for Elizabethan audiences, but were
> modern directors to adopt a less conventional approach, that fact alone
> would not compel the conclusion that the work was obscene.[87]

Responding to the government's argument that virtual pornography encourages the market in pornography and whets the appetite of people
who exploit children, Justice Kennedy continued, "The mere tendency

of speech to encourage unlawful acts is not sufficient reason for banning. First Amendment freedoms are most in danger when the government seeks to control thought or to justify its laws for that impermissible end. The right to think is the beginning of freedom, and speech must be protected from the government because speech is the beginning of thought."[88]

But the marketplace kept creeping into Justice Kennedy's argument, and he jumped into the unknown when he said, "If virtual images were identical to illegal child pornography, the illegal images would be driven from the market by the indistinguishable substitutes. Few pornographers would risk prosecution by abusing real children if fictional, computerized images would suffice."[89] In other words, since the portrayal of virtual sex does not involve actual children and therefore has no victims and causes no collateral damage to society, it is a protected form of speech, however perverted, and its correction, if necessary, should be left to the marketplace rather than to thinking people. This was certainly not the intention of the founding fathers when they wrote, "Congress shall make no law."

It would seem that the marketplace has gobbled up freedom of speech in the United States, making congressional efforts to regulate sexual expression on the Internet through the Communications Decency Act, the Child Online Protection Act, the Children's Internet Protection Act, and the Child Pornography Protection Act a futile exercise. It is possible that the courts have come to realize that since it is humanly impossible to police the Internet, the only way to control its chaotic freedom is through filtering technology, the solution that authoritarian regimes such as China, Saudi Arabia, Singapore, and Myanmar have been successfully implementing. If filtering technology can save American children from finding, for example, a breast, penis, or vagina on the Internet, it can also protect the Chinese from Falun Gong, the Dalai Lama, and democracy. Filtering technology is nonmoral, and in the era of burgeoning social networks, YouTube, MySpace, and the like, there is a growing international market for software that restores social controls.

Chapter Seven

How Does the Ring
of Freedom Sound to You?

As wireless chips are embedded into more and more devices, the Internet is becoming ubiquitous and virtual and more powerful with grid computing. But whether digital mobility, speed, and virtualization will enhance our freedom—for example, if we were able to surf the Web through eyeglasses, listen to music on iPod-like devices, conduct legal or clinical trials, buy and sell real estate, or enter into a loving relationship in the metaverse of Second Life—will depend on how we frame that freedom.

The Internet collapses space and time and brings people together whether they like it or not. Since every human activity, from pornography to the most complex mathematical hypothesis, is nothing but information, whatever takes place in the analog world can be turned into a digital stream, a flow of binary mathematical code that can instantly be distributed globally through the Internet, thus extending the reach of human communication, creating new forms of reality, and increasing choices for good and evil.[1] Messages turned into digital data can be changed into value-added information and transformed into predictive intelligence for human behavior in order to enhance commerce, national security, or other social or political activities. Digital books, music files, or terrorist messages are indistinguishable as they converge and surge through cyberspace.[2] They can be transmitted and distributed instantly and more inexpensively than in the analog world. Convergence is value-enhanced transformation.

Convergence, instantaneity, and interactivity make the Internet a powerful medium of spontaneous communication, a self-generating medium
that mines and transforms messages, creates choices, and enhances freedom at the same time that it strengthens controls. Since the traditional
communications media, including books, television, newspapers, magazines, radio, music, and interpersonal communication, are converging on
the Internet as a multimedia stream anyone can plug into (e.g., Wikipedia
and YouTube), their systemic power increases manyfold. A drop becomes an ocean and loses its identity. If to be free is to make choices
freely, the Internet paradoxically enhances and diminishes our abilities to
do so because every choice creates possibilities for surveillance and control.[3] MySpace and Second Life are glass houses.

Harold Innis wrote in *The Bias of Communication* that a new medium
of communication creates a specific cultural bias and has "an important
influence on the dissemination of knowledge over space and time and it
becomes necessary to study its characteristics in order to appraise its influence in its cultural setting."[4] Although people in ancient times tried
to abridge space and time by sending messages, especially in wartime,
through drums and smoke signals, only after the invention of the telegraph was it possible to think about communication in terms of transmission rather than transportation. Like goods, messages in the pretele-
graph era were transported from place to place at the speed made
possible by the best transportation system of the time; for example, consider the train and the pony express or the legendary messenger who ran
forty-two kilometers from the battlefield of Marathon to Athens to an-
nounce the Athenian victory over the Persians in 490 BC. The telegraph
altered the geography-based metaphor of communication, which ceased
to be synonymous with transportation, and as new technologies—telephone, radio, and television—developed in the early part of the twentieth century, communication became increasingly liberated from the
constraints of geography. Computer networks and the Internet have further altered our perspective on space and time. A networked organization or an individual with instant messaging and e-mail has a different
sense of space and time than people of the predigital era. The keyboard
is the door to cyberspace, and all who enter are simultaneously in a
world that is synchronous and asynchronous, one that gives a greater illusion of freedom and control than the real world.[5] The illusion breaks
up when users realize that they're under some impersonal digital sur-

veillance built into the system and are no longer free. The right not to be seen by others is freedom, but once people exercise their freedom to surf and search for information in order to choose a song or buy a product on e-Bay or Amazon, they lose their freedom to be left alone. Freedom to access takes away the freedom to be private. A person in the metaverse of cyberspace is as free as he would be in a casino in Las Vegas, where every activity is watched, analyzed, and stored in a digital database for crime prevention and for future reference in order to generate more business. The transition from an era of physically transported messages to that of electronic transmission to today's spontaneous Internet communication and telepresence is a thumbnail history of communication.

Although we are far from realizing the world of cyberspace conjured up by William Gibson in his science fiction novel *Neuromancer* (1984), a world in which computers directly connect with the human nervous system to create a virtual world, cyberspace is evolving into a multidimensional universe in which we can do things that are as real as they are in physical space. The presence of others in a virtual environment created by networked communication is called telepresence; it allows, for example, geographically dispersed surgeons to control a robot remotely to perform a surgical operation or a group of scientists and engineers in Bangalore and Boston to design a prototype, solve a mathematical puzzle, or crack a mystery. Increasingly, there is "the pervasive recognition that a new and decentered spatiality has arisen that exists parallel to, but outside of, the geographic topography of experiential reality."[6] As more and more people use and experience telepresence and work in virtualization for personal and professional reasons, they will begin to feel unsure about what is real and what is virtual, what is private and what is public. Uncertainty creates anxiety and diminishes the sense of freedom—we may sometimes feel this way when we conduct a transaction on the Internet. As we expand our choices in the postmodern digital world, invisible chains increasingly chain us. Freedom gained seems to be freedom lost, as if some invisible force were controlling freedom.

THE CHOICES WE MAKE

"Everything's horrible, I want to die. Who will die with me?" That was the message in a Japanese Internet chat room after four people between the ages of nineteen and thirty were found to have committed suicide in a car parked on a riverbank on the island of Hokkaido in northern Japan in 2005.[7] For the group, the message had become a transformative and collaborative experience made possible by the Internet.

In 1973, when two young computer scientists, Vinton G. Cerf and Robert E. Kahn, came up with the revolutionary idea of making different isolated computers talk to each other through a common language, the Transmission Control Protocol/Internet Protocol, they could not have foreseen the whole new world of possibilities and dangers that would eventually open up. Of course, many more people made the Internet possible, eventually, looking at the bright side, making Bangalore, for example, a global outsourcing hub. Cerf and Kahn could not have anticipated that the Internet would become a global driving force for good—and evil—in our lives. E-trading, e-pornography, e-surveillance, e-death, and who knows what else are all in store for us, for the simple reason that the Internet is a self-generating and self-sustaining system of decentralized networks open to all, saints and sinners, which makes further networking and sharing inevitable.

Consider the case of a twenty-six-year-old man, Gerald Krein from Klamath Falls, Oregon, who tried enticing thirty-two women in chat rooms to commit mass suicide on Valentine's Day. Klamath County Sheriff Tim Evinger told the media, "The common theme is that these were women who were vulnerable, who were depressed. He invited them to engage in certain sexual acts with him—and they were to hang themselves naked from a beam in his house." Krein might have used his Webcam to netcast the event to the world—thirty-two women hanging naked from a roof beam. Sexual asphyxiation is a most extreme sexual act, and in a land of choices, including death by choice—Oregon allows physician-assisted suicide as a solution for terminal patients—it would have probably created a minor shock, then would have been shrugged off. Strangers encounter each other in cyberspace, form temporary communions, then disappear.

A Canadian woman who saw the message titled "Suicide Ideology" in a chat room and learned, to her horror, that another chat room woman intended to kill not only herself but her two children promptly informed the police. At least thirty-one women had agreed to participate in the mass suicide, Krein told the police investigators upon his arrest. Chat room records show that Krein had been networking with women to solicit suicide since 2000.[8]

Using extreme acts culminating in suicide in response to deep depression may also explain why some people blow themselves up for jihad, which Internet networking and sharing makes so blindingly enticing. Dying alone is terrible, but it becomes easier when people die together, as the Japanese group found. The Internet provides opportunities for togetherness with faceless strangers, providing the freedom to act and freedom from loneliness, as YouTube, MySpace, Facebook, and other social networks demonstrate.

FRAMING FREEDOM

Not long ago, Barry Schwartz, Hazel Rose Markus, and Alana Conner Snibbe carried out a survey about how Americans frame freedom. They found that despite the popular view that more choices mean greater freedom, for most people, too many choices can be confusing, not liberating. According to their research, most middle- and upper-class Americans equate freedom with choice because their incomes enable them to show off their personalities and preferences through the choices they make. On the other hand, for most Americans whose incomes are limited, "being free is less about making choices that reflect their uniqueness and mastery and more about being left alone, with their personality, integrity and well-being intact." This is a profoundly valid statement that is equally applicable to people in the Arab-Muslim world, who cherish a different idea of freedom.[9]

Being away from scrutiny, being able to preserve one's sense of self and dignity, is freedom for some people, especially the poor, while for others, especially the affluent, freedom is "choice, control and self-expression." For celebrities, freedom means publicity that enhances their brand value, which ironically reduces their privacy, their freedom to be left alone. More choices, as Schwartz, Markus, and Snibbe conclude, do not necessarily increase one's freedom:

> In part because of the higher social status of middle-class Americans, the equation of freedom with choice is the one most loudly broadcast. Every corner of life is now rife with choices, as well as with talk of the control and self-expression that choosing imparts. But is this middle-class conception of freedom the "right" one? Empirical evidence suggests that we should be careful what we wish for. Americans are increasingly

overwhelmed by all these choices. We feel less free now than when we
had fewer choices, and we show it in our behavior.[10]

From freedom as the ability to make choices, Schwartz, Markus, and
Snibbe make a bold and imaginative leap and question the wisdom of
the current foreign policy initiative of making other people as free as
Americans are, burdening them with choices that they cannot handle.
"To govern well, both at home and abroad," they admonish, "Americans would be wise to listen to how freedom rings in different cultural
contexts."[11] To see ourselves as others see us may be a gift from God,
as the Scottish poet Robert Burns (1759–1796) said, but to insist on seeing others as we see ourselves is morally wrong and can be diplomati-
cally disastrous.

The kind of freedom most Americans have in mind, that is, freedom
to make consumer choices, is not necessarily the foundation of democracy; otherwise, the Chinese, with the increasing number of consumer
choices available to them due to rising prosperity, would have taken definitive steps toward a democratic form of government. It is a leap of
faith that abundant consumer choices will lead to a burning desire for
political choices and democracy, one for which there is no historical evidence. Abundant consumer choices can become an opiate of the masses
and an alternative to freedom—one that is no different from the Muslim
paradise teeming with virgins for deserving jihadists—an alternative to
political freedom that China is trying to create for its teeming masses.

Democracy as a dynamic system of power sharing and accountable
government is essentially a latter-day European and American experience, even though its origins lie in ancient Greece and it is successfully
practiced today in many countries, including India, the largest functioning democracy. A free marketplace of goods does not create a free marketplace of ideas, and even if the two are found together, they do not
necessarily give rise to democracy.

Traditions of public reasoning and discussion, apart from those of ancient Greece and Rome, have been prevalent in other cultures, for example, Persia, India, and China. To build the Great Wall of China, the
Taj Mahal, the pyramids, or nuclear weapons, some form of discussion
and public reasoning, however limited, is essential. "Since traditions of
public reasoning," wrote the economist and Nobel laureate Amartya Sen
optimistically, "can be found in nearly all countries, modern democracy
can build on the dialogic part of the common human inheritance."[12] Pro-

fessor Sen cites the great Mughal emperor Akbar for "codifying minority rights, including religious freedom for all, along with championing regular discussions between followers of Islam, Hinduism, Jainism, Judaism, Zoroastrianism and other beliefs (including atheism)."[13] But Akbar was a shrewd, autocratic ruler of a non-Muslim country who used the policy of acceptance and tolerance as a political strategy rather than as a plan to limit and share his power in a democratic sense, and everything the emperor accomplished died with him because he left no political structure to sustain tolerance and public discussion. Democratic freedom cannot survive without a dynamic institutional structure.

The tradition of public reasoning, freedom of expression, and tolerance is good for its own sake and is also an indispensable tool of democracy, but in various degrees, you may have these tools of democracy without democracy. Democracy as a system of structuring and distributing power, a system that allows the exercise of power but can be revoked in case of corruption, misuse, and inefficiency, is essentially a European and American experiment, though in the twentieth century, it has been adopted in other countries successfully, such as Japan and India. A democratic system distributes power according to functions, and no one is vested with power more than is absolutely necessary to perform the duties for which he or she is accountable. Voting is a method of electing and vesting some people with limited power and subsequently holding them answerable through periodic elections; freedom of expression, apart from being a social good, is an indispensable tool to bring about change.

Power held too long and without accountability tends to corrupt, to paraphrase Lord Acton; therefore, it must be limited in its exercise and duration, whether the form of government is parliamentary or presidential. The idea that the absolute rights of the ruler must be limited began with the Magna Carta of 1215, which enjoined the king (King John) to give up some of his rights, respect legal procedures, and be bound by the law. In this way, it gave rise to the development of constitutional law as the source of political power and democratic government, which makes authorities answerable for how they use power.[14] It was this accountability for power in a parliamentary constitutional monarchy that obligated Robert Clive, the East India Company's civil servant who laid the foundation of British rule in India, to face an impeachment trial in the British House of Commons in 1773, although he was exonerated of all charges. Winston Churchill, who along with Franklin D. Roosevelt is credited with winning the war for democracy over fascism, was defeated in the 1945 general elections by Clement Attlee and the Labour Party because the British electorate believed that a war prime minister was not good enough to rebuild the peace and create prosperity. Richard Nixon was forced to resign the U.S. presidency because he misused power and acted as if he were above the law. Bill Clinton faced impeachment for perjury and obstruction of justice, though the Senate finally acquitted him. It is unimaginable that such events might happen in China, for example, even if the country became a full market economy.

In *Hamdan v. Rumsfeld,* the U.S. Supreme Court majority ruled that President George W. Bush had exceeded his authority in establishing military tribunals for the trial of al Qaeda detainees held in the U.S. military base at Guantánamo Bay, Cuba. The decision came as a rude shock to the administration, which had been behaving and acting on the assumption that, during wartime, the president needs exceptional authority to protect the nation, may even bypass Congress if necessary, and is, in fact, above the law. In his concurring opinion, Justice Anthony Kennedy said, "Concentration of power puts personal liberty in peril of arbitrary action of officials, an incursion the Constitution's three-part system is designed to avoid."[15] The landmark decision reaffirmed the dynamic power structure of checks and balances that has institutionalized fundamental freedoms in the United States, regardless of how freedom rings for each individual.

Accountability of power is at the heart of democracy, which cannot be accomplished unless the people have freedom in all its manifestations, including freedom of action, thought, expression, access, choice, and movement, as well as freedom from fear and freedom to be left alone, among many, many others.

A culture of public reasoning and discussion, freedom of choice, free press, and voting are indispensable building blocks of democracy, but it is possible to have the building blocks without democracy—witness China, where 9 to 10 percent economic growth for the last three decades could not have occurred without some freedoms, or Pakistan, a country with a very healthy and robust free press but no democracy.

If the sound of freedom has a different ring for people living in different cultures, how much freedom does a person need? How much freedom can a person handle? The question echoes the theme of Leo Tolstoy's story "How Much Land Does a Man Need?" It is the story of a peasant who, in his greed for more land, loses his life and is buried in a six-by-two-foot plot of land. This is in fact the moral of the story and beautifully drives home the answer to the question posed in the title. One of Russia's greatest playwrights and Tolstoy's contemporary, Anton Chekhov, commented on the story, "It is a common saying that a man needs only six feet of earth. But six feet is what a corpse needs, not a man. . . . Man needs not six feet of earth, not a farm, but the whole globe, all of nature, where unhindered he can display all the capacities and peculiarities of his free spirit."[16] This is the kind of illusion of freedom most of us have, and one is likely to be carried away by the eloquence of a great writer like Chekhov, but the question is, How do you sustain so much freedom in a society?

STRUCTURE OF FREEDOM

Freedom in its various manifestations—whether it is freedom to do or not to do something, freedom of choice, access, expression, faith, or movement, or freedom from wants and fears—cannot exist without a dynamic system of structures that constitutionally balance each other and create space for freedom to emerge.

Commenting that freedom arises from a law-governed society, David Boaz of the Cato Institute said in a 2004 panel discussion, "I think the Constitution has maintained a largely free society for a long time and that is an accomplishment. The question is whether there would be something that would maintain a freer society for longer."[17] No, no, no! Having a constitution does not make a society free. Many countries are governed by constitutions but lack most freedoms that the people of democratic societies enjoy. The Soviet Union and Saddam Hussein's Iraq had constitutions, but no one has ever accused those nations of having freedom. Today, China and Iran are constitutionally governed, but the sounds of freedom are so different in these countries from what they are in Western-style democracies. A constitution alone cannot guarantee freedom. Nor can all the trappings of democracy—voting, elections, and a free press—guarantee that freedom in all its forms will flourish.

Freedom arises from a contrapuntal dynamic system that limits the exercise of power, be it that of the government or a corporation or some religious institution, especially one that subscribes to a doctrine of absolutism. The U.S. Constitution has maintained our freedom because at its heart is a fortress of negativity, the Bill of Rights, which asks the government not to trespass into the lives of the people unless there is a compelling reason in accordance with the law as interpreted by the Court. The tripartite system of dynamic power relations among the co-equals predicated upon a system of mutually interacting fundamental rights has sustained freedom in American society for a long time. Freedom is an emergence, the function of a system in dynamic equilibrium that makes the powerful accountable to the people who made them powerful.

Freedom is a virtue in itself and the ultimate human goal. Equally importantly, however, freedom is a vital and indispensable necessity, the source of a society's creativity, inventions, and innovations, which are needed to generate a perpetual cycle of self-renewal. Freedom fuels the generative mechanism of a society, but it has to be structured and shaped so that it does not become self-destructive. It is very legitimate to ask, Where does freedom come from? How much freedom does a person need? How much freedom can a person handle?

HEALTHY DISTRUST

People in power must never be fully trusted and should be subjected to the dictum, "Trust but verify." A free society does not need heroes. For freedom to flourish and function, there must be a degree of separation between the government and the people. A healthy distrust between the people and the government is essential for an objective evaluation of the government's exercise of power. If the Bill of Rights were done away with, the United States might devolve into a most horrendous form of Latin American or Middle Eastern dictatorship. The Bill of Rights, one of the most important documents ever written to limit the power of the government in support of human rights and individual freedom, tells the government, in a definitive language reminiscent of the Ten Commandments (Thou shall not . . . Congress shall not . . .) that the rights of the people are inviolable and the power of the government is limited. The government nonetheless always looks for ways and means to bypass the secular commandments on which American civilization has been founded. The Bill of Rights, consisting of the first ten amendments to the

U.S. Constitution, was ratified in 1791, and after the Civil War, it was extended to the states through the Fourteenth Amendment, popularly known as the due process and equal protection amendment.

The tripartite system of checks and balances among the coequals in power, reinforced by the Bill of Rights, has been an enduring structure of freedom and a perennial source of strength and self-renewal for American society. Instead of trusting random and ephemeral democratic impulses, the founding fathers created a durable democratic structure that has allowed freedom to exist and flourish in the United States for more than two centuries. The rhetoric of freedom is essential to emotionalize people about the value of freedom, but freedom cannot exist in a vacuum. Freedom emerges not only from the mind of humans but also from the institutions they build.

Though rising from the backdrop of European turmoil and the tyranny of both the church and the state, the American political system was founded on a pessimistic, but realistic, view of human nature, because if people are not vigilant, there is always the danger that a society will slip into chaos and barbarism, into the "heart of darkness."[18]

The epic journey from Athenian to American democracy has not passed from one beautiful valley to another. Beneath the veneer of culture and civilization, there is a raging sea of anarchy, death, and destruction into which any society can plunge if continuous watchfulness is not exercised to protect and renew the institutional structures that made freedom possible. There is no movement from dawn to everbrighter dawn, but only the possibility of reversal, of being sucked into barbarism, as great writers like Joseph Conrad have reminded us.

A recent example of a society that has slipped back into barbarism is Afghanistan, where a few years ago the Taliban regime felt so threatened by the country's own beautiful, rich, ancient heritage that it destroyed ancient Buddha statues; two of them, carved into the Bamiyan cliffs in the heart of the Hindukush Mountains, were the tallest in the world. Worse than destroying their own cultural heritage was their attempt to enslave women totally, reducing them to household chattel. Cambodia under Pol Pot and Germany under the Nazis are other examples of how some societies with constitutions but without checks and balances can sink into ethnic cleansing and barbarism.

Constitutionally guaranteed and enforceable fundamental rights not only check the excesses of executive power but also have the potential to stop the abuse of unbridled technology used by businesses for self-aggrandizement. Many people naively believe that digital technology and the Internet will free them from the tyranny of the government and big business. This is not necessarily so. The Nazis used IBM punch card technology to profile and separate Jews from others to hasten their tragic journey to concentration camps and the Final Solution.[19] No collaboration was intended, and it would be wrong to jump to any conclusions because hitherto no clear and convincing evidence has emerged from the catacombs of history that there was indeed a conspiracy of silence or cooperation. But there are cases of religious institutions, banks, and multinational corporations that have used surveillance technology and committed heinous crimes against humanity, and they must be held accountable. The U.S. government used census data to identify and herd Japanese Americans into internment camps during World War II. Out of fear, the American people shut the door on freedom and forgot the Bill of Rights. All systems, whether political, cultural, legal, or religious, tend to dissipate and degenerate unless they are periodically challenged and renewed. Every new technology becomes another source of power, which creates a potential for its misuse; therefore, accountability—how power is used and why—is a necessary condition for the self-renewal of a free society, without which privacy and other freedoms cannot exist.

DEGREES OF FREEDOM

Today, we measure a nation's openness and its worth in terms of quality of life, in terms of the free flow of information within its borders and the degree of freedom it gives its people to speak, write, and publish without fear of punishment. The ancient Greek worldview that "man is the measure of everything" (Protagoras) is essentially the mode of consciousness today in the West and in liberal democracies like India and Japan. In many ways, the West boasts of its liberal traditions, but it has not always been so liberal, as we briefly noted in the case of the barbaric treatment of the publisher William Twyn in seventeenth-century England.

Great civilizations in the past were built without the general public's enjoying freedom of thought and expression, but none could have flourished for long without internal debate, however controlled it might have been. Great literatures and works of art could not have been created without freedom of expression, however limited by the laws of the land. In the city-state of Athens in the fifth century BC, where democracy flourished for a while, those in power could not tolerate Socrates's questioning of authority. He was charged with corrupting the morals of the youth and sentenced to exile or death by drinking hemlock. But his death was not in vain because his method of discovering truth through dialectic and open discussion has become a universal legacy. Freedom and democracy did not last long in ancient Greece because seeking truth for the sake of truth was deemed a destabilizing social factor at a time when Greece was facing external threats.

Neither the ancient Greeks and Romans nor the medieval church thought it worthwhile to let freedom trickle down to the masses because it would not have served any function of the state or those in power, the secular or religious authorities. Only in the seventeenth and eighteenth centuries, during the Enlightenment in the West, when people began to move away from both God and the church and looked at themselves in a different light, as individuals capable of reason and self-government, did the idea of freedom as a natural right for self-fulfillment and a necessary condition for independent decision making begin to be widely accepted. But because having all natural rights and no authority would create chaos, individuals must give up some rights and powers to the government to create a society of laws that protects their freedom, as the seventeenth-century philosopher John Locke propounded.[20] Locke's seminal thinking had a profound effect upon the founders of the American Constitution, who limited the government's powers by erecting the greatest structural barriers to tyranny ever known to mankind: a tripartite government of coequals in power, with further checks and balances embodied in the Bill of Rights, based on the notion of private property as sacrosanct. In this strong, yet flexible, constitutional framework, freedom of speech has gradually established its preferred position in relation to other rights over the last two hundred years of American history, particularly since the beginning of the twentieth century.

Half a century before John Locke, the great English poet John Milton eloquently argued that freedom of speech can ultimately create the essential social good and enable society to discover truth through the dialectic of the marketplace. Because the Puritans, who ruled England in the mid-sixteenth century, had imposed censorship, Milton (ironically, he was the official censor of the Puritan government) essentially argued against prepublication censorship. Milton wanted the dissolution of his troubled marriage and wanted freedom to argue in support of divorce, which the Puritan government did not allow. Although his motives in arguing against prior restraint were not altruistic, he elevated his personal problem into a fascinating treatise justifying the establishment of a marketplace of ideas:

> And though all the winds of doctrine were let loose to play upon the earth, so Truth be in the field, we do injuriously, by licensing and prohibiting, to misdoubt her strength. Let her and Falsehood grapple; who ever knew Truth put to the worse, in a free and open encounter.[21]

Human beings seek fulfillment in multitudinous ways, but there are some, like Patrick Henry (1736–1799), who would say, "Give me liberty or give me death." Self-expression in any form is central to the idea of being free for some, while others seek self-fulfillment through religious expression. Though freedom as a "natural right" appeared first during the Enlightenment, in the United States it began to be accepted as the quintessence of humanity and a guiding political value with the signing of the Declaration of Independence and outbreak of the American Revolution in 1776. In time, freedom of speech and expression came to be perceived as a generative instrument of social good, as well as a means of public participation in political and economic decision making.

As a source of individual self-fulfillment and an instrument for enhancing political participation and commerce, free speech has been transformed from a philosophical "natural right" to a necessary condition for living in the postmodern digital age, an age where all values are ultimately tested in the competition of the marketplace.

The marketplace, a metaphor for freedom, is a forum for conversation, exchange, competition, and innovation in the postmodern digital society. A self-governing society with a free marketplace at its center cannot do without free speech and the free flow of ideas. The reinterpretation of Milton that it is only through the "clash of ideas in the marketplace" that truth can be reached has made free speech an indispensable tool of innovation. Innovation is an indispensable social good for a self-renewing society. For the fullest development of individual potential, freedom in its myriad forms is an absolute necessity for which we need contrarian institutional structures in dynamic equilibrium that create space for freedom.

However, despite the centrality of secularism in the free marketplace of ideas to the postmodern digital age in the United States, other societies have different systems of core values, the dome of attraction that bonds them, defines their cultures, and determines the degrees of digital freedom they need for which they use filters and gateways as well as social and market controls. No wonder the drums of freedom sound so different in China, Iran, Saudi Arabia, and the United States.

Notes

PREFACE

1. Yochai Benkler, *The Wealth of Networks: How Social Production Transforms Markets and Freedom* (New Haven, CT: Yale University Press, 2006).
2. Lawrence Lessig, *Code and Other Laws of Cyberspace* (New York: Basic Books, 1999).
3. Lawrence Lessig, *The Future of Ideas: The Fate of the Commons in a Connected World* (New York: Random House, 2001), *Free Culture: How Big Media Uses Technology and Law to Lock Down Culture and Control Creativity* (New York: Penguin Press, 2004), and *Code: Version 2.0* (New York: Basic Books, 2006).
4. Siva Vaidhyanathan, *Copyrights and Copywrongs: The Rise of Intellectual Property and How It Threatens Creativity* (New York: New York University Press, 2001).

CHAPTER 1

1. Chris Lucas, "Self-Organizing Systems FAQ," Version 2.99, July 2006, www.calresco.org/sos/sosfaq.htm (accessed 16 March 2007). Erich Jantsch, *The Self-organizing Universe* (New York: Pergamon Press, 1980).
2. Lawrence Lessig, *Code and Other Laws of Cyberspace* (New York: Basic Books, 2000). "The invisible hand, through commerce, is constructing an architecture that perfects control—an architecture that makes possible highly efficient regulation" (6).

3. Katie Dean, "Instant Messaging Grows Up," *Wired News* 8, no. 1 (January 2000), www.wired.com/news/culture/0,1284,33736,00.html (accessed 24 March 2007).

4. For a discussion of the strength of the weakest bond, see Cosma Rohilla Shalizi, "Notebooks: Dissipative Structures," Center for the Study of Complex Systems, University of Michigan, January 28, 1997, www.cscs.umich.edu/~cr-shalizi/notebooks/dissipative-structures.html (accessed 24 March 2007).

5. Joe Ashbrook Nickell, "Teens Enter Virgin Territory," *Wired News* 6, no. 7 (July 15, 1998), http://wired-vig.wired.com/news/culture/0,1284,13761,00 .html (accessed 15 July 1998).

6. "Giving Birth Online," *Time*, June 17, 1998, www.time.com/time/community/transcripts/chattr061798.html (accessed 17 June 1998).

7. "Poets are the unacknowledged legislators of the world," is a quotation from P. B. Shelley, "A Defense of Poetry," in *English Essays: From Sir Philip Sidney to Macaulay,* ed. Charles W. Eliot (New York: P. F. Collier and Son, 1909), www.bartleby.com/27/23.html (accessed 2 March 2000).

8. "R.I.P. Jennicam," BBC News, http://news.bbc.co.uk/1/hi/magazine/3360063.stm (accessed 24 March 2007).

9. Jock Gill, "Universal Access Session," The Harvard Conference on the Internet and Society, May 1996, www.jockgill.com/presentations/may96/harv0596.html (accessed 16 March 2007).

10. See Nicholas Lemann, "Amateur Hour," *New Yorker*, August 7 and 14, 2006, 44–49.

11. Samuel P. Huntington, *The Clash of Civilizations and the Remaking of World Order* (New York: Touchstone, 1996).

CHAPTER 2

1. Nicholas Negroponte, *Being Digital* (New York: Alfred A. Knopf, 1995).

2. Neil Postman, *Amusing Ourselves to Death: Public Discourse in the Age of Show Business* (New York: Penguin, 1985).

3. "Technorealism Overview," Technorealism, www.technorealism.org (accessed 12 May 2002).

4. Sherry Turkle, *The Second Self: Computers and the Human Spirit* (Cambridge, MA: MIT Press, 1984).

5. *Smart House*, Internet Movie Database, www.imdb.com/title/tt0192618/ (accessed 5 June 2000).

6. Lawrence Lessig, *Code and Other Laws of Cyberspace* (New York: Basic Books, 1999).

7. Iver Peterson, "'Smart Guns' Setting Off Debate: How Smart Will They Really Be?" *New York Times*, http://select.nytimes.com/gst/abstract.html?res= F30914FA3B590C718EDDA90994D0494D81 (accessed 22 October 1998).

8. William J. Mitchell, *e-topia: Urban Life, Jim—But Not As We Know It* (Cambridge, MA: MIT Press, 1999).

9. Conversation in January 2000 with Sudarshan Khurana, an engineer with Parsons, a global engineering and construction company.

10. Wade Roush, "Social Machines," *Technology Review*, August 2005, 44–53. Roush says, "A leading example in the United States is the Georgia Tech Broadband Institute Residential Laboratory, a three-story home outfitted with people-tracking sensors, gesture-sensitive remote controls, and other widgets. Part of the Aware Home Research Initiative funded by Hewlett-Packard, Intel, Motorola, and the National Science Foundation, the Residential Laboratory is a classic instance of computing research that starts with a perceived need—assisting the elderly with complex, information-intensive tasks, for example—and invents gadgets and software that supposedly addresses the need" (47).

11. Dana Kirsch, "The Sentic Mouse: Developing a Tool for Measuring Emotional Valence," Affective Computing, http://affect.media.mit.edu (accessed 3 September 2002).

12. Mitchell, *e-topia*, 46.

13. Joseph S. Nye Jr., *The Paradox of American Power* (New York: Oxford University Press, 2002).

14. For more on Internet2, see www.internet2.edu/about/ (accessed 5 January 2004).

15. Quoted in Michael M. Waldrop, "Grid Computing," *Technology Review*, May 2002. Also see "Grid Computing Planet: Opinions: Interviews: IBM's Wladawsky-Berger Explains Grid Computing," IBM, www-1.ibm.com/linux/grid/InterviewwithIBM.doc (accessed 22 March 2002).

16. Douglas M. Towns, "Employment Non-Compete Agreements in the Internet Revolution," Jones Day, www1.jonesday.com/lawyers/bio.aspx?attorney ID=S2034&op=publications (accessed 30 August 2000).

17. Barbara Pierce, "NFB Sues AOL," National Federation of the Blind, www.nfb.org/Images/nfb/Publications/bm/bm99/bm991201.htm (accessed 24 March 2007)). NFB, according to the website, is "A comprehensive evaluation, demonstration, and training technology center, which contains over $2 million worth of speech and Braille technology," www.nfb.org/nfb/Technology_Center.asp?SnID=274941 (accessed 24 March 2007). The case was settled out of court on 26 July 2000.

18. *Carparts Distribution Center v. Automotive Wholesalers Association of New England*, http://harp.org/carparts.txt (accessed 17 March 2007).

19. See the Internet Freedom Act (1998), College of Business, San Jose State University, www.cob.sjsu.edu/nellen_a/ITFA.html (accessed 15 March 2000). The act was extended twice by Congress and is scheduled to expire in November 2007.

20. See H.R. 1684: Internet Tax Nondiscrimination Act of 2005, GovTrack.us, www.govtrack.us/congress/bill.xpd?tab=main&bill=h109-1684 (accessed 5 June 2003).

21. Forrester, "U.S. eCommerce: 2005 to 2010," Forrester, September 14, 2005, www.forrester.com/Research/Document/Excerpt/0,7211,37626,00.html (accessed 14 September 2005).

22. CNN, "Georgia Killer's Notes Show a Troubled Man," CNN, July 30, 1999, www.cnn.com/US/9907/30/atlanta.shooting.06 (accessed 30 July 1999).

23. "The purchase and sale (or the short sale and cover) of the same *security* on the same day, also called day trade." See "Daylight Trade," InvestorWords.com, www.investorwords.com/1289/daylight_trade.html (accessed 4 November 2000). A Toronto woman told me on December 5, 2000, "I just pressed the key, and the next moment I realized that I had lost $90,000."

24. Andrew Shapiro, *The Control Revolution* (New York: Century Foundation, 1999). Also see Michael Barbaro and Tom Zeller, "A Face Is Exposed for AOL Searcher No 441749," *New York Times,* August 9, 2006, 1.

25. For a discussion of "creative destruction," see Joseph A. Schumpeter, *Capitalism, Socialism and Democracy* (1942; rprt. New York: Harper, 1975), 82–85.

26. "Digital Manipulation Code of Ethics," Statement of Principle, National Press Photographers Association, www.nppa.org/professional_development/business_practices/digitalethics.html (accessed 17 December 2004).

27. Richard Rosecrance, *The Rise of the Virtual State: Wealth and Power in the Coming Century* (New York: Basic Books, 1999).

28. For a definition of synergy, see AskOxford.com, www.askoxford.com/asktheexperts/faq/aboutgrammar/synergy (accessed 24 March 2007).

29. "Murdoch's Web Gambit," *Businessweek*, October 10, 2005, www.businessweek.com/magazine/content/05_41/b3954044.htm (accessed 10 October 2005).

30. Tim Berners-Lee, *Weaving the Web: The Original Design and the Destiny of the World Wide Web* (New York: HarperCollins, 1999).

31. "TerraGrid: Unleashing Clustered Computing," HPC Wire, www.taborcommunications.com/hpcwire/hpcwireWWW/03/1118/106440.html (accessed 2 February 2004).

32. See "Larry Smarr's Recent Talks," Jacobs School of Engineering, University of California, San Diego, www.jacobsschool.ucsd.edu/~lsmarr/talks/index.html (accessed 27 December 2001).

33. "IBM's Irving Wladawsky-Berger Explains Grid Computing," IBM, www-1.ibm.com/linux/grid/InterviewwithIBM.doc (accessed 31 March 2002).

34. Kenneth Chang, "Scientists Report They Have Made Robot That Makes Its Own Robots," *New York Times*, August 31, 2000, http://select.nytimes.com/gst/abstract.html?res=F00615FA3F540C728FDDA10894D8404482 (accessed 31 August 2000).

35. Bill Joy, "Why the Future Doesn't Need Us," *Wired*, April 2000, http://wired-vig.wired.com/wired/archive/8.04/joy.html?pg=1&topic=& topic_set= (accessed 20 April 2000).

CHAPTER 3

1. For discussion of privacy, see, for example, Arthur R. Miller, *The Assault on Privacy* (Ann Arbor: University of Michigan Press, 1971); Don R. Pember, *Privacy and the Press* (Seattle: University of Washington Press, 1974); Alan Westin, *Privacy and Freedom* (New York: Atheneum, 1967); Whitfield Diffie and Susan Landau, *Privacy on the Line* (Cambridge, MA: MIT Press, 1998).

2. See Vance Packard, *The Naked Society* (New York: David McKay and Co., 1964).

3. *Olmstead v. United States*, 277 U.S. 438, 48 S.Ct. 564 (1928).

4. Samuel Warren and Louis D. Brandeis, "The Right to Privacy," *Harvard Law Review* 4 (1980), 196. Quoted in Harold L. Nelson and Dwight L. Teeter Jr., *Law of Mass Communications*, 5th ed. (New York: Foundation Press, 1986), 205.

5. Pember, *Privacy and the Press*, 5.

6. Robert Ellis Smith, ""Ben Franklin's Web Site: Privacy and Curiosity from Plymouth Rock to the Internet," *Privacy Journal*, 2000, 19. Quoted in Patrick Marshall, "Privacy under Attack," *CQ Researcher* 11, no. 23 (June 2001).

7. *Briscoe v. Reader's Digest Association*, 4 Cal.3d 529, 93 Cal.Rpt. 866, 483 2d 34, 36–37 (1971). Also see Nelson and Teeter, *Law of Mass Communications*, 205.

8. Pember, *Privacy and the Press*, 50–51. See *DeMay v. Roberts*, 46 Mich. 160 (1881), for the first privacy case of its kind. A Michigan woman won the case for invasion of privacy against the doctor who had allowed his unqualified assistant to be present during the delivery. Available at History of Privacy, http://historyofprivacy.net/DeMay.htm (accessed 15 June 2004). Also see Patricia Mell, "Seeking Shade in a Land of Perpetual Sunlight: Privacy as Property in the Electronic Wilderness," University of California, Berkeley, School

of Law, www.law.berkeley.edu/journals/btlj/articles/vol11/Mell/html/text.html (accessed 23 January 2005).

9. For example, see Jeffrey Blyth, "*People* Hopes 5m Thanks to Brad, Angelina and Shiloh," Online Press Gazette, www.pressgazette.co.uk/article/120606/people_hopes_for_5m_sales_thanks_to_brad_angelina_and_shiloh (accessed 12 June 2006).

10. William Pitt (1708–1778) said, "The poorest man may in his cottage bid defiance to all the forces of the Crown. It may be frail; its roof may shake; the winds may blow through it; the storms may enter—but the King of England cannot enter; all his forces dare not cross the threshold of the ruined tenement." Available at Bartleby.com, www.bartleby.com/73/861.html (accessed 31 May 2004).

11. See "Thermology," www.thermology.com (accessed 2 January 2003).

12. "Thermal Camera Helps Keep Track of Cattle Health," *St. Louis Post-Dispatch*, December 24, 2000, D9. Also see "Thermal Cameras Find Drug Factories," Find Articles, findarticles.com/p/articles/mi_qn4153/is_20060323/ai_n16163546 (accessed 18 March 2007).

13. Charles Lane, "Justices Hear Oregon Case on High-Tech Surveillance," *Washington Post*, February 21, 2001, LexisNexis, http://web.lexis-nexis.com/universe/docum (accessed 23 March 2001); ProQuest, http://proquest.umi.com.library.norwich.edu/pqdweb?index=0&did=68898330&SrchMode=1&sid=1&Fmt=3&VInst=PROD&VType=PQD&RQT=309&VName=PQD&TS=1174853933&clientId=55007 (accessed 24 March 2007).

14. *Kyllo v. United States*, No. 99-8508. Justices Antonin Scalia and John Paul Stevens, "Excerpts from Supreme Court Ruling on Limits on Surveillance Technology," *New York Times*, June 12, 2001, A21, http://query.nytimes.com/gst/fullpage.html?res=9B0CE3D8133EF931A25755C0A9679C8B63&n=Top%2FReference%2FTimes%20Topics%2FOrganizations%2FS%2FSupreme%20Court%20 (accessed 18 March 2007).

15. "Excerpts."

16. "Excerpts."

17. "Excerpts."

18. Regarding e-mail as court evidence, see Nicholas Varchaver, "The Perils of E-mail," *Fortune*, February 17, 2003, 96.

19. "Carnivore Diagnostic Tool," Center for Democracy and Technology, www.cdt.org/security/carnivore/000724fbi.shtml (accessed 24 August 2000).

20. "Carnivore Diagnostic Tool."

21. "Internet and Data Interception Capabilities Developed by the FBI," Statement for the Record, U.S. House of Representatives, the Committee on the Judiciary, Subcommittee on the Constitution, 07/24/2000, Laboratory Division Assistant Director Dr. Donald M. Kerr. Available at Center for Democracy and Technology, www.cdt.org/security/carnivore/000724kerr.htm (accessed 24 August 2000).

22. "U.K. E-mail Snooping Bill Passed," CNN, July 28, 2000, http:// archives.cnn.com/2000/TECH/computing/07/28/uk.surveillance.idg/index.htm l (accessed 2 August 2000).

23. Daniel Henninger, "Back to Sanity? National Security Is a Video Game for Politicians," *Wall Street Journal,* May 26, 2006, www.opinionjournal .com/columnists/dhenninger/?id=110008429 (accessed 26 May 2006).

24. Tom Clancy's *Net-Force* (1999), available on DVD.

25. U.S. Foreign Policy Agenda, "Information Warfare Threat Demands More Attention on All Sides: An Interview with Senator Jon Kyl," *USIA Electronic Journal* 3, no. 4 (November 1998), www.fas.org/irp/threat/cyber/docs/ usia/pj48kyle.htm (accessed 3 December 2000).

26. "Cyber Attack: Is the Nation at Risk?" Senate Governmental Affairs Committee, June 24, 1998, Hearing, eStrategy, www.estrategy.gov/documents/ cyber.htm (accessed 1 July 1998).

27. "Climber Saved by Text Message," *BBC News*, http://news.bbc .co.uk/1/hi/wales/1341955.stm (accessed 1 June 2001).

28. *Bartnicki et al. v. Vopper*, a.k.a. Williams et al., Nos. 99-1687 and 99-1728, Supreme Court of the United States, 2001 U.S. Lexis 3815, 69 U.S.L.W. 4323.

29. *Bartnicki et al. v. Vopper.*

30. Federal Wiretap Act 1986.

31. *Bartnicki et al. v. Vopper.*

32. *New York Times Co. v. United States*, 403 U.S. 713, 91 S.Ct. 2140, 29 L.Ed.2d 822 (1971).

In *New York Times v. Sullivan*, 376 U.S. 254, 84 S.Ct. 710, 11 L.Ed.2d 686 (1964), the U.S. Supreme Court ruled that public officials must prove "actual malice," that is, reckless disregard of the truth, in order to win damages for defamation.

33. Jonathan Alter, "Prime Time Revolution," *Newsweek*, January 8, 1990, 25.

34. Mel Ayton, "The Truth about J. Edgar Hoover," *Crime Magazine: An Encyclopedia of Crime*, http://crimemagazine.com/05/jedgarhoover,0719-5.htm (accessed 5 July 2005).

35. *Katz v. United States*, 389 U.S. 353 (1967).

36. Ralph L. Holsinger and John Paul Dilts, *Media Law*, 3rd ed. (New York: McGraw Hill, 1994), 223.

37. The Rebecca Schaeffer account is available at http://movies2 .nytimes.com/gst/movies/filmography.html?p_id=63535&mod=bio (accessed 4 April 2003).

38 *Reno v. Condon* (98-1464) 528 U.S. 141 (2000), 155 F.3d 453, reversed, available at the Legal Information Institute, Cornell Law School, www.law.cornell .edu/supct/html/98-1464.ZO.html.

39. *Reno v. Condon.*

40. *Saenz v. Roe,* (98-97) 526 U.S. 489 (1999), 134 F.3d 1400, affirmed.

41. "U.S. Ruling Makes Libel in Cyberspace Punishable," *New York Times,* December 18, 2000, www.uri-geller.com/libel_net.htm (accessed 24 March 2007).

42. David Briggs, "Pastoral Counseling Offered Online," *Times-Picayune* (New Orleans), November 25, 2000, http://web.lexis-nexis.com/universe/document (accessed 1 December 2000).

43. Briggs, "Pastoral Counseling."

44. Norma Love, "Lawmaker Resigns over Stance on Killing Police," *Associated Press,* January 11, 2001, www.findarticles.com/p/articles/mi_qn4196/is_20010111/ai_n10680984 (accessed 23 January 2003).

45. Associated Press, "Child Porn Victim Says Justice Did Not Protect Him," *USA Today,* April 5, 2006, www.usatoday.com/news/washington/2006-04-04-child-porn_x.htm. Joshua Brockman, "Child Sex as Internet Fare, through Eyes of a Victim," *New York Times,* April 5, 2006, www.nytimes.com/2006/04/05/washington/05porn.html?ex=1145073600&en=7442a813236 57d17&ei=5070 (accessed 10 April 2006).

46. *Viereck v. United States,* 318 U.S. 236, 251 (1943), University of Miami School of Law, www.law.miami.edu/froomkin/articles/oceanno.htm (accessed 17 November 2000).

47. Martin Gottlieb, "Pattern Emerges in Bomber's Tract," *New York Times,* August 2, 1995, A1.

48. *Church of Scientology vs. the Net,* Computer Professionals for Social Responsibility, www.cpsr.org/prevsite/cpsr/nii/cyber-rights/web/current-scientology.html (accessed 4 January 2005).

49. *NAACP v. Alabama ex rel. Patterson,* 357 U.S. 449 (1958).

50. *McIntyre v. Ohio Elections Commission,* 514 U.S. 334 (1995).

51. *Bank of Boston v. Bellotti,* 435 U.S. 765, 792n32 (1978).

52. A. Michael Froomkin, "Flood Control on the Information Ocean: Living with Anonymity, Digital Cash, and Distributed Databases," *Pittsburgh Journal of Law and Commerce* 395 (1996), www.law.miami.edu/~froomkin/articles/oceanno.htm (accessed 20 June 2000).

53. Jodi Upton, "U-M Medical Records End Up on Web," *Detroit News,* February 12, 1999, A1, ProQuest, http://proquest.umi.com/pqdweb?did=38911000&sid=1&Fmt=2&clientId=55007&RQT=309&VName=PQD. One of the most recent cases of data loss due to negligence was the loss of sensitive information on 26.5 million veterans by a Veterans Affairs data analyst, who took a laptop home. The data were subsequently recovered with apparently no damage done. Johanna Neuman, "Veterans Chief Is Grilled over Stolen Data," *Los Angeles Times,* May 26, 2006, A5, ProQuest, http://proquest.umi.com/pqdweb?did=1042252291&sid=2&Fmt=3&clientId=55007&RQT=309&VName=PQD (accessed 24 March 2007).

54. "Medical Privacy Opinion Polls," Epic.org, www.epic.org/privacy/medical/poll.htm (accessed 22 December 2005).

55. Twila Brase, "M.D. Confidential: The Government Is Intruding on Patients' Right to Privacy," *Minnesota Physician*, February 1999, www.cchconline.org/publications/mpppriv.php3 (accessed 2 March 1999).

56. The Bush administration put the medical privacy rules, the Standards for Privacy of Individually Identifiable Health Information, which were to take effect on April 14, 2001, on hold. See Health and Human Services, www.hhs.gov/ocr/hipaa/finalmaster.html (accessed 1 May 2002).

57. A Florida woman sued Walgreen's and Eli Lilly and Co. in July 2002 for disclosing information about her antidepressant prescription drug Prozac. The company had earlier released the e-mail addresses of six hundred Prozac patients but promised the Federal Trade Commission to strengthen its online security system. An HIV-positive man in Pennsylvania lost his job when the company learned that he was taking AIDS drugs. A North Carolina woman lost her job because her employer learned that she had an incurable genetic disease. See "Protecting the Privacy of Patients' Health Information," U.S. Department of Human and Health Services, "Fact Sheet," www.hhs.gov/news/facts/privacy.html (accessed 18 March 2007)

58. "Protecting Electronic Health Information," Committee on Maintaining Privacy and Security in Health Care, Applications of the National Information Infrastructure, National Research Council (Washington, DC, 1977). See also Simon Garfinkel, *Database Nation* (Sebastopol, CA: O'Reilly and Associates, Inc., 2000), 150. For the latest health data theft, see Bob Sullivan, "VA Loses Another Computer with Personal Information," MSNBC, http://msnbc.msn.com/id/14232678 (accessed 30 July 2006).

CHAPTER 4

1. Consider the dissenting views of U.S. Supreme Court Justice Louis D. Brandeis in *Olmstead v. U.S.* (1928), which, in the course of time, the Court came to accept as a legal norm: "The maker of our Constitution . . . sought to protect Americans in their beliefs, their thoughts, their emotions and their sensations. They conferred as against the Government, the right to be let alone—the most comprehensive for the rights of man and the right most valued by civilized men."

2. In *Katz v. United States*, 389 U.S. 347 (1967), the Court ruled that wiretapping a telephone conversation without probable cause violates a person's Fourth Amendment rights.

3. See Raymond Williams, *Marxism and Literature* (New York: Oxford University Press, 1977), 110. "It [hegemony] is a whole body of practices and

expectations, over the whole of living: our senses and assignments of energy, our shaping perceptions of ourselves and our world. It is a lived system of meanings and values—constitutive and constituting—which as they are experienced as practices appear as reciprocally confirming. It thus constitutes a sense of reality for most people in the society, a sense of absolute."

4. In *United States v. Miller*, 425 U.S. 435, 442 (1976), the Court ruled that the bank depositor could not claim reasonable expectation of privacy because that information was voluntarily disclosed to the bank, and the government was not required to meet Fourth Amendment standards.

5. *Smith v. Maryland*, 442 U.S. 735, 743, 746 (1979).

6. Title III of the Omnibus Crime Control and Safe Street Act of 1968 (Wiretap Act), 18 U.S.C. 2510–20 (1968).

7. 18 U.S.C. 2510–21 (1986).

8. 18 U.S.C. 2701–11 (1994).

9. *U.S. v. Bianco*, 998 F.2d 1112 (1993); *U.S. v. Gayton*, 74 F. 3d 545 (1996).

10. *United States v. Charbonneau*, 979 F. Supp. 1177 (S.D. Ohio 1997).

11. *United States v. Maxwell*, 45 M.J. 406 (1996).

12. "Carnivore Diagnostic Tool," Center for Democracy and Technology, www.cdt.org/security/carnivore/000724fbi.shtml (accessed 5 February 2002).

13. Information about wiretapping is available at http://epic.org/privacy/wiretap (accessed 3 March 2003).

14. USPA, Sections 214 and 216.

15. USPA, Section 210.

16. USPA, Sections 202 and 209.

17. President Bush signed the renewed Patriot Act on March 9, 2006. See Federation of American Scientists, www.fas.org/sgp/crs/intel/RL33332.pdf (accessed 3 April 2006).

18. USPA, Section 217.

19. See "USA Patriot Act Improvement and Reauthorization Act of 2005: A Legal Analysis," Federation of American Scientists, www.fas.org/sgp/crs/intel/RL33332.pdf (accessed 2 January 2006).

20. USPA, Section 212.

21. USPA, Section 211.

22. USPA, Sections 219 and 220.

23. USPA, Section 206.

24. Justice Louis Brandeis in *Olmstead v. United States*, 277 U.S. 438 (1928).

25. According to Jason Erb of the Council on American-Islamic Relations, "It starts to erode some of the trust and good will that exists in these institutions if you're afraid they have been infiltrated by an undercover agent." Quoted in Adam Liptak, "Traces of Terror: News Analysis; Changing the Standard," *New York Times*, May 31, 2002, A1.

26. Susan Stellin, "Terror's Confounding Online Trail," *New York Times*, March 28, 2002, http://query.nytimes.com/gst/fullpage.html?res=9D07EFD 8103BF93BA15750C0A9649C8B63 (accessed 1 April 2002): "For all the sophisticated electronic tools the United States government has at its investigative disposal, tracking the activities of suspected terrorist groups online has proved to be not unlike the search for Osama bin Laden and his operatives on the ground. In essence, even against a superior arsenal of technology, there are still plenty of ways for terrorists to avoid detection."

27. Shannon P. Duffy, "Digging through Old E-mails Not Wiretap Violation," *Legal Intelligencer*, March 28, 2001, http://web.lexis-nexis.com/universe/document? (accessed 9 August 2002).

28. Rachel Konrad, "Blue Cross Insurer Says Pornographic Messages Violated Company Rules," *Detroit Free Press*, September 30, 1999, http://personal.lig.bellsouth.net/B/r/Brucej11/smut.htm (accessed 22 March 2007).

29. In May 1999, Harvard University fired the dean of the Divinity School because links to pornography were found on his school computer. See Kevin W. Bowyer, "Pornography on the Dean's PC: An Ethics and Computing Case Study," University of Notre Dame, www.nd.edu/~kwb/nsf-ufe/PornOnDeansPC.pdf (accessed 30 May 2000).

30. "2001 AMA Survey: Workplace Monitoring and Surveillance," American Management Association, June 6, 2001, www.amanet.org/research/pdfs/ems_short2001.pdf (accessed 30 May 2002).

31. Patrick Marshall, "Privacy under Attack," *CQ Researcher* 2, no. 23 (June 15, 2001): 507.

32. Ann Bartow, "Our Data, Ourselves: Privacy, Propertization, and Gender," *University of San Francisco Law Review* (Summer 2000), http://eon.law.harvard.edu/privacy/OurDataOurSelves(Bartow).htm (accessed 19 March 2002).

33. See Oxygen Media at www.oxygen.com. A typical day's topic, according to the Website, might include "What does he think? Or does even he think? Get the scoop on our 'he said/she said' message boards, where women can find out a man's opinion . . . not that it matters." The site includes such topics as health and fitness, careers, family and pregnancy, lifestyle, relationships, and sex, all topics of great interest to women.

34. The Women's Network, www.ivillage.com, offers similar services as Oxygen Media. Both portals collect personal information from women. When a woman responds to polls (e.g., What kind of birth control do you use?), she leaves plenty of personal information behind. In some sense, iVillage and Oxygen Media are similar to *Cosmopolitan*.

35. Web bugs can do much more than cookies, which only keep track of surfers when they log on to a site. Web bugs can clandestinely steal data files from a hard drive without the knowledge of the user. Newer bugs with

different functions are being developed. Yahoo! used a Web beacon, a single-pixel electronic image, to keep track of the visitors to its portal and "to conduct research on behalf of certain partners on their websites and also for auditing purposes." Website meters yield complete visitor information, including the geographic location, time, date, browser, and computer used.

36. "Public Workshop on Consumer Privacy on the Global Information Infrastructure," FTC Staff Report, December 1996, Section I, Federal Trade Commission, www.ftc.gov/reports/privacy/privacy1.htm (accessed 4 June 2000).

37. Stuart F. Brown, "Building America's Anti-terror Machine," *Fortune*, July 22, 2002, 102.

38. FTC Staff Report, December 1996.

39. DoubleClick created a furor in 2000 when it announced that it would combine consumer information stored on its offline databases with clickstream data it collected from surfers with the tracking technology. As surfers move from site to site, they leave a digital trail of mouse droppings, which can be turned into a data pattern that can be mined for further information for marketing and intelligence purposes. But the company backed off its plans because of the pressure from the public and federal regulators.

40. The Video Privacy Protection Act of 1988, Family Educational Rights and Privacy Act of 1974, Driver's Privacy Protection Act of 1994, Fair Credit Reporting Act of 1999, Privacy Act of 1974, and Electronic Communications Privacy Act of 1986 deal with some aspect of personal information privacy.

41. Paul M. Schwartz and Joel R. Reidenberg, *Data Privacy Law: A Study of United States Data Protection* (Dayton, OH: Lexis Law Publishing, Michie, 1996). Quoted in Karl D. Belgum, "Who Leads at Half-time? Three Conflicting Visions of Internet Privacy Policy," *Richmond Journal of Law and Technology* 6, no. 1, Symposium 1999, http://cyber.law.harvard.edu/privacy/Who LeadsatHalftime(Belgum).htm (accessed 19 March 2002).

42. Arthur Miller, *The Assault on Privacy* (Ann Arbor: University of Michigan Press, 1971), 20. Alan Westin, *Privacy and Freedom* (New York: Atheneum, 1967), 160–63.

43. Belgum, "Who Leads at Half-time."

44. Jeffrey Reimen, "Driving to the Panopticon, A Philosophical Exploration of the Risks to Privacy Posed by the Highway Technology of the Future," *Computer and Hightech Law Journal* 11 (1995): 27, 28n2. Quoted in Belgum, "Who Leads at Half-time."

45. Henry S. Perrit Jr., "Regulatory Models for Protecting Privacy in the Internet," in *Privacy and Self-Regulation in the Information Age* (U.S. Department of Commerce, 1997), 107–109, www.ntia.doc.gov/reports/privacy/privacy_rpt.htm (accessed 5 July 1998).

46. See Charles J. Sykes, *The End of Privacy* (New York: St. Martin's Press, 1999), 29: "Every piece of information has a price. Your salary and consumer

credit card report can be obtained from an information broker for $75; your stock, bond and mutual-funds records for $200. For $450, a broker can obtain your credit card number; for $80 to $200 they can put their hands on your telephone records. Your personal medical history for the last ten years is for sale for $400. Not all of this information can be obtained ethically or even legally. But it can be obtained." A company like ChoicePoint, Inc., provides subscription-based access to its databases regarding taxpayers' assets, driving records, addresses, phone numbers, and other personal information.

47. The White House, "A Framework for Global Electronic Commerce," July 1, 1997, Technology Administration, www.technology.gov/digeconomy/framewrk.htm (accessed 15 August 1998): E-commerce "on the GII [Global Information Infrastructure] will thrive only if the privacy rights of individuals are balanced with the benefits associated with the free flow of information." Also see Peter Swire and Robert E. Litan, *None of Your Business: World Data Flows, Electronic Commerce, and the European Privacy Directive* (New York: Brookings Institution Press, 1998).

48. Information is synthesized from various sources. See Department of Commerce, *Elements of Effective Self-Regulation for the Protection of Privacy and Questions Related to Online Privacy*, No. 980422102-8102-01, www.ntia.doc.gov/ntiahome/privacy/6_5_98fedreg.htm (accessed 5 July 1998).

49. EU Directive 95/46/EC of the European Parliament and of the Council of 24 October 1995 on the protection of individuals with regard to the processing of personal data and the free movement of such data. Available at the Center for Democracy and Technology, www.cdt.org/privacy/eudirective/EU_Directive_.html (accessed 22 February 2003).

50. Dan Froomkin, "Deciphering Encryption," *Washington Post*, May 8, 1998, www.washingtonpost.com/wp-srv/politics/special/encryption/encryption.htm (accessed 7 March 1999).

51. Whitfield Diffie and Susan Landau, *Privacy on the Line* (Cambridge, MA: MIT Press, 1998), 13.

53. Lehnert, Wendy, *Web 101,* 642.

54. Since a key is a string of zeros and ones, a computer uses its power to test each possible string of zeros and ones until the right one is found. For example, a 10-bit key has 1,024 possibilities, and a 56-bit key has 72,057,594,037,927,936 possibilities. This looks like an extremely large number, but it's not for computers. See Lehnert, *Web 101*, 629.

55. See "TerraGrid: Unleashing Clustered Computing," HPC Wire, www.taborcommunications.com/hpcwire/hpcwireWWW/03/1118/106440.htm l (accessed 5 January 2005).

56. Clipper chip technology would have enabled backdoor entry into encrypted messages. Since it would have been integrated into the software of

approved manufacturers "in the form of tamper-resistant integrated circuits," the chips would have created a trapdoor for allowing intelligence and law-enforcement authorities to intercept traffic when needed. See Diffie and Landau, *Privacy on the Line*, 211.

57. Diffie and Landau, *Privacy on the Line*, 277: "Before public-key technology, cryptography always required centralized facilities to manufacture equipment and keys, a feature particularly compatible with top-down organization of the military. By contrast, public-key cryptography was developed to support the interactions of businesses in a community of equals."

58. The Church Committee (the Senate Select Committee to Study Governmental Operations with Respect to Intelligence Activities) concluded, "The Government has often undertaken the secret surveillance of citizens on the basis of their political beliefs, even when those beliefs posed no threat of violence or illegal acts on the behalf of a foreign power. The Government, primarily operating through secret informants, but also using other intrusive techniques such as wiretaps, microphones, 'bugs,' surreptitious mail opening, and break-ins, has swept in a vast amount of information about the personal lives, views and associations of American citizens" (USS 94d, 5, quoted in Diffie and Landau, *Privacy on the Line*, 178). The Church Committee also concluded, "Unjustified investigations of political expression and dissent can have a debilitating effect upon our political system. When people see this can happen, they become wary of associating with groups that disagree with the government and more wary of what they say or write. The impact is to undermine the effectiveness of popular self-government" (USS 101, 1, quoted in Diffie and Landau, *Privacy on the Line*, 147–48).

59. These are the words of Judge William Campbell of U.S. District Court in Northern Illinois before Congress in 1965. Quoted in Diffie and Landau, *Privacy on the Line*, 170.

60. Quoted in Diffie and Landau, *Privacy on the Line*, 103, 107.

61. Platform for Privacy Preferences (P3P) Project, www.W3.org/P3P (accessed 16 April 2002).

62. For the EU's directive, see the Center for Democracy and Technology, www.cdt.org/privacy/eudirective/EU_Directive_.html (accessed 5 May 2001).

63. "P3P and Privacy: An Update for the Privacy Community," Center for Democracy and Technology, March 28, 2000, www.cdt.org/privacy/pet/p3pprivacy.shtml (accessed 16 April 2002).

64. "Choose to Hold TRUSTe Members to Higher Standards," TRUSTe, www.truste.org/pvr.php?page=complaint (accessed 16 April 2002).

65. "Choose to Hold TRUSTe Members to Higher Standards," TRUSTe, www.truste.org/pvr.php?page=complaint.

66. Karen Coyle, "A Response to 'P3P and Privacy: An Update for the Privacy Community' by the Center for Democracy and Technology," Karen Coyle, May 2000, www.kcoyle.net/response.html (accessed 16 May 2000).

67. William F. Adkinson Jr., Jeffrey A. Eisenach, and Thomas. M. Lenard, "Privacy Online: A Report of the Information Practices and Policies of Commercial Websites," Progress and Freedom Foundation, Washington, DC, March 2002.

68. David Krane, "Privacy Leadership Initiative: Privacy Notices Research," Harris Interactive, December 2001, www.ftc.gov/bcp/workshops/glb/presentations/krane.pdf (accessed 4 January 2002).

CHAPTER 5

1. The Chinese were the first to invent clay movable-type printing in 1041, but unlike in Europe, printing had little effect on Chinese society. See Mary Bellis, "Johannes Gutenberg and the Printing Press," About:Inventors, http://inventors.about.com/library/inventors/blJohannesGutenberg.htm (accessed 30 June 2003).

2. See "Can This Pandemic Be Stopped?" Small Business School, http://smallbusinessschool.org/webapp/sbs/sbs/index.jsp?page=http%3A%2F%2Fsmallbusinessschool.org%2Fwebapp%2Fsbs%2FMedia%2Fhomepage.jsp (accessed 12 July 2004).

3. Even the United States has limited power of persuasion when it comes to dealing with a country like China. See, for example, John Schwartz, "Black Market for Software Is Sidestepping Export Controls," *New York Times*, December 2, 2002, C1.

4. Bruce A. Lehman, "Intellectual Property and the National Information Infrastructure: The Report of the Working Group on Intellectual Property Rights," November 1995, U.S. Patent and Trademark Office, www.uspto.gov/web/offices/com/doc/ipnii (accessed 3 September 2005).

5. James Boyle, "Intellectual Property Policy Online: A Young Person's Guide," Duke Law, www.law.duke.edu/boylesite/joltart.htm (accessed 23 May 2004).

6. Lehman, "Intellectual Property."

7. The Sonny Bono Copyright Term Extension Act (1998) is available from the Association of Research Libraries website at www.arl.org/info/frn/copy/extension.html (accessed 30 June 2002).

8. For a discussion of Walt Disney's creativity, see Lawrence Lessig at Stanford Law School: Lawrence Lessig, www.lessig.org/blog/archives/000788.shtml (accessed 28 August 2004).

9. Lawrence Lessig, "Free Culture: Lawrence Lessig Keynote from OSCON 2002," O'Reilly Policy Devcenter, www.oreillynet.com/pub/a/policy/2002/08/15/lessig.html?page=1 (accessed 14 August 2002).

10. Information about the Utah Lighthouse Ministry is available from their website at www.utlm.org (accessed 15 August 2002).

11. 75 F.Supp.2d 1290, 1999 U.S. Dist. LEXIS 19103, 53 U.S.P.Q.2D (BNA) 1425, Copy. L. Rep. (CCH) P28,013 (1999). Available from the University of Houston Law Center, www.law.uh.edu/faculty/cjoyce/copyright/release10/IntRes.html (accessed 22 July 2001).

12. *Zacchini v. Scripps-Howard Broadcasting Co.*, 433 U.S. 562 (1977).

13. *John W. Carson v. Here's Johnny Portable Toilets, Inc.*, 698 F.2d 831, 1 218 USPQ 1, 9 Media L. Rep. 1153 (1982).

14. *Allen v. Men's World Outlet, Inc.*, 679 F. Supp. 360 (S.D.N.Y. 1988).

15. See Jonathan Rosenoer, *Cyber Law* (New York: Springer-Verlag, 1997), 62–75. Also see Emily Eakin, "The Censor and the Artist: A Murky Border," *New York Times*, November 26, 2002, B3, in which she says, "Copyright holders, not government, turn out to be a new kind of gatekeeper."

16. *UMG Recordings, Inc. v. MP3.com, Inc.*, 92 F. Supp.2d 349 (S.D.N.Y. 2000). For a news report about the case, visit CNN at http://archives.cnn.com/2000/LAW/09/06/mp3.lawsuit.01 (accessed 2 October 2000). The original MP3.com was shut down on December 2, 2003, and reappeared as a music site with the same name operated by CNET Networks.

17. For a news report about the Metallica case, visit CNN at http://archives.cnn.com/2001/LAW/02/12/napster.decision.05/index.html (accessed 1 March 2001). For the Grokster decision, visit News.com at http://news.com.com/Supreme+Court+rules+against+file+swapping/2100-1030_3-5764135.html (accessed 1 July 2005). For *Universal v. Sharman* (Kazaa case), visit the Australian Copyright Council at www.copyright.org.au/news/newsbytopic/recentcases/U26147/ (accessed 1 July 2005). Also see Eric Pfanner, "Kazaa Deal Ends P2P Suites," *International Herald Tribune*, July 28, 2006, www.iht.com/articles/2006/07/27/yourmoney/music.php (accessed 22 March 2007).

18. Information about DeCSS is available at Gallery of CSS Descramblers, www.cs.cmu.edu/~dst/DeCSS/Gallery (accessed 27 March 2007).

19. *2600: The Hacker Quarterly* is available at www.2600.com (accessed 23 July 2005).

20. Andrea L. Foster, "The Making of a Policy Gadfly," *Chronicle of Higher Education*, November 29, 2002, A27.

21. Information about the Secure Digital Management Initiative (SDMI) is available from the Berkman Center for Internet and Society at Harvard Law School, http://cyber.law.harvard.edu/property00/MP3/sdmi.html. Edward Felton presented his research findings in 2001 at the USENIX security conference without being threatened by any lawsuit by the SDMI group, which seems to have been inactive since 2001.

22. Carl S. Kaplan, "Experts Say Decision Could Undermine Online Journalists," *New York Times,* December 24, 2001, www.nytimes.com/2001/12/14/technology/circuits/14CYBERLAW.html?ex=1146888000&en=7dcfd0d91b5d1f28&ei=5070 (accessed 25 December 2001); also see *MPAA v. 2600*, cy-

ber.law.harvard.edu/openlaw/DVD/NY/appeals/010501-oral-arg.html (accessed 25 December 2003).

23. *Feist Publications, Inc. v. Rural Telephone Service Co.*, 111 S.Ct. 1282, 1294 (1991). Justice Sandra Day O'Connor wrote, "The original requirement is not particularly stringent. A compiler may settle upon a selection or arrangement that others have used; novelty is not required. Originality requires only that the author make the selection or arrangement independently (i.e., without copying the selection or arrangement from another work) and that it display some minimum level of creativity."

24. *Twentieth Century Music Corp. v. Aikin*, 422 U.S. 151, 156 (1975).

25. John D. Zelezny, *Communications Law*, 4th ed. (Belmont, CA: Wadsworth, 2004), 331.

26. *Sony Corporation of America v. Universal City Studios, Inc.*, 464 U.S. 417 (1984). Also see *Basic Books, Inc. v. Kinko's Graphic Corp.*, 758 F. Supp. 1552 (1991), in which the U.S. District Court for the Southern District of New York ruled that the "potential for widespread copyright infringement by Kinko's and other commercial copiers is great," and unauthorized copying, even though for educational proposes, could damage the market value of copyrighted materials.

27. *Harper & Row, Publisher, Inc. v. Nation Enterprises*, 471 U.S. 539, 105 S.Ct. 2218, 85 L.Ed.2d 588, 11 Med. L. Rptr. 1969 (1985).

28. In *Harper & Row* (1985), Justice O'Connor wrote, "In view of the First Amendment protection already embodied in the Copyright's distinction between copyrightable expression and uncopyrightable facts and ideas, and the latitude for scholarship and comment traditionally afforded by fair use, we see no warrant for expanding the doctrine of fair use to create what amounts to a public figure exception to copyright. Whether verbatim copying from a public figure's manuscript in a given case is or is not fair must be judged according to the traditional equities of fair use."

29. *Salinger v. Random House, Inc.*, 811 F.2d 90 (2d Cir. 1987), cert denied, 484 U.S. 890 (1987). In 1980s Ian Hamilton, a literary critic with the London *Sunday Times,* was working on the biography of J. D. Salinger, who had been living the life of a recluse and shunned all media attention. When the author came to know about the project, he sued, asking for exclusive rights to his unpublished letters. The court sided with him.

30 *Campbell v. Acuff-Rose Music*, 510 U.S. 569 (1994). In *Hustler v. Falwell*, 485 U.S. 46 (1988), a unanimous Supreme Court gave strong protection to social criticism in the form of parody and satire.

31. See, for example, Melville B. Nimmer, "Does Copyright Abridge the First Amendment Guarantees of Free Speech and Press?" *UCLA Law Review* 17 (1969–1970): 1181; Paul Goldstein, "Copyright and the First Amendment," *Columbia Law Review* 70 (1970): 983–1057; Thomas L. Tedford and Dale A.

Herbeck, *Freedom of Speech in the United States*, 4th ed. (State College, PA: Strata Publishing, Inc.), 343.

32. For example, the song "Dora Dean" by Bert Williams (published by Broder and Schlam, New York) was denied copyright in 1898 for the line "She is the hottest thing you ever seen," and another popular song that hinted at immoral sexual behavior, "Rum and Coca-Cola" (1940s), was initially denied copyright, though copyright was subsequently granted, according to Norman A. Palumbo Jr., "Obscenity and Copyright: An Illustration Past and Future?" *South Texas Law Journal* 22 (1982): 87–101. Quoted in Tedford and Herbeck, *Freedom of Speech,* 345.

33. *Rosemont Enterprises v. Random House*, 366 F.2d 203 (2d Cir. 1966; cert denied).

34. Nimmer, "Does Copyright Abridge the First Amendment Guarantees of Free Speech and Press?"

35. *Time, Inc. v. Bernard Geis Associates*, 293 F. Supp. 130 (S.D.N.Y. 1968).

36. *Triangle Publications v. Knight Ridder Newspapers*, 445 F. Supp. 875 (S.D. Fla. 1978).

37. *International News Service v. Associated Press*, 248 U.S. 215 (1918). The International News Service was enjoined from lifting news from Associated Press sources until "the commercial value" of stories had passed.

38. See "Terms and Conditions of Use for the martindale.com Web Site," Martindale.com, www.martindale.com/xp/Martindale/Site_Info/terms.xml (accessed 22 June 2002).

39. Mark Stefik, "Trusted Systems," *Scientific American*, www.sciamdigital.com/index.cfm?fa=Products.ViewIssuePreview&ARTICLEID_CHAR=7D4812C7-BC39-4203-9B17-CD9B547ABEF (accessed 22 March 2007).

Lawrence Lessig, "The Code Is the Law," *Industry Standard* 9 (April 1999), www.lessig.org/content/standard/0,1902,4165,00.html (accessed 24 February 2002).

40. Lessig, "The Code Is the Law."

41. John Perry Barlow, "The Economy of Ideas," *Wired*, www.wired.com/wired/archive/2.03/economy.ideas_pr.html (accessed 23 June 2000).

42. Stefik, "Trusted Systems."

43. In "Open Source Definition" in *Open Sources: Voices from the Open Source Revolution*, ed. Chris DiBiona, Sam Ockman, and Mark Stone (Sebastopol, CA: O'Reilly and Associates, Inc., 1999), Bruce Perens of the Open Source Initiative (OSI) writes that open source is more than access to the source code. The definition of open source includes the following: (1) Redistribution must be free, which means "The license shall not restrict any party from selling or giving away the software as a component of an aggregate program from several different sources." (2) Both the source code and compiled form (human unreadable code) must be made available. (3) Derived works and modifications

are permitted. (4) The integrity of the author's source code must be maintained so that modifications will not be confused with the original code. (5) There is no discrimination against any persons or groups. (6) There is no discrimination against any field of endeavor. (7) Distribution of licenses should be automatic, without requiring a signature. (8) Licenses must not be limited to a particular product. (9) Licenses must not impose restrictions on any other software. (10) Licenses must not be made dependent upon any particular technology or interface style. These rules are available from the Open Source website at www.opensource.org (accessed 23 November 2002).

44. Paul Wallich, "The Best Things in Cyberspace Are Free," *Scientific American*, March 1999, www.sciamdigital.com/index.cfm?fa=Products.View IssuePreview&ARTICLEID_CHAR=025C3CCC-9BBA-4CA5-9E3E-1C91F424A61 (accessed 22 March 2007).

45. See Perens, "The Open Source Definition."

46. Perens, "The Open Source Definition."

CHAPTER 6

1. The Securities and Exchange Commission (SEC) and other U.S. law-enforcement authorities do not have jurisdiction in other countries where many online crimes originate, but they have posted warnings on their websites about the kinds of Internet investment frauds usually committed. Some of them, according to SEC, include the following:

- The "Pump and Dump" scam: The investment expert or analyst claims to have "inside information" about a forthcoming deal or development that will raise stock prices and urges the reader to buy, buy, buy. When the prices go skyward, the fraudster sells and disappears, and the stock prices hit rock bottom.
- The pyramid scheme: This old trick in new garb promises that an investor can turn "$5 into $60,000 in just three to six weeks."
- The "Risk-Free" scheme: Investors are urged to put their money into a cutting-edge technological development with guaranteed returns or money back.

In addition to the above, there are, of course, offshore frauds, originating in Russia, Nigeria, and other countries living on the edge of chaos.

There is no end to the ingenuity and creativity of Internet fraudsters, who come up with scams faster than the SEC can catalog them, but no one beats a gentleman e-mailer who wanted me to accept his client's $42 million and save a besieged South African family from the cruel and unjust hands of the government of free South Africa. To learn how to protect yourself from Internet fraud, visit the SEC website at www.sec.gov/investor/pubs/cyberfraud.htm (accessed 2 May 2005). Also see Mitchell Zuckoff, "The Perfect Mark," *New Yorker*, May 15, 2006, 36–42.

2. Jan H. Samoriski, "Unsolicited Commercial E-Mail, the Internet and the First Amendment: Another Free Speech Showdown in Cyberspace," *Journal of Broadcasting and Electronic Media* 43, no. 4 (Fall 1999). Also see "A Brief History of Junk," JunkBuster, www.junkbusters.com/variety.html (accessed 3 June 2003). For the dubious status of spam, that is, how by masquerading as free speech it trumps privacy, see Reynolds Holding, "A Spammer's Revenge," *Time*, January 15, 2007.

3. *Juno Online Services, L.P. v. Scott Allen Export Sales*, No. 97 Civ. 8694, S.D.N.Y (1997).

4. *Rowan v. U.S. Post Office Dept.*, 397 U.S. 728 (1970), 737.

5. *Douglas K. Snow v. Daniel Doherty*, U.S. District Court for the Northern District of Indiana, South Bend District Court, Civil Action File No 3:97-CV0635 (RM).

6. For example, see *America Online, Inc. v. Cyber Promotions, Inc.*, 948 F. Supp. 436 (1996) (E.D. PA).

7. *Cyber Promotions, Inc. v. America Online, Inc.*, C.A. No. 96-2486, 1996 WL 565818 (E.D. Pa. Sept. 5, 1996) (temporary restraining order), *rev'd* (3d Cir. Sept. 20, 1996), *partial summary judgment granted*, 948 F. Supp. 436 (E.D. Pa. Nov. 4, 1996) (on First Amendment issues), *reconsideration denied*, 948 F. Supp. 436, 447 (Dec. 20, 1996), *temporary restraining order denied*, 948 F. Supp. 456 (E.D. Pa. Nov. 26, 1996) (on antitrust claim), *settlement entered* (E.D. Pa. Feb. 4, 1997).

8. Refusing to accept advertising can touch antitrust law. For example, see the landmark case *Lorain Journal Co. v. United States*. The publisher of a newspaper with a 97 percent circulation in the area advised the advertising department to cancel the advertising contract of any Lorain business that advertised with the radio station broadcasting from a nearby town. The purpose, the U.S. Supreme Court ruled, was to drive the radio station out of business by using its monopoly power and therefore violated the law. 342 U.S. 143, 72 S.Ct. 181, 96 L.Ed. 162, 1 Media L. Reporter, 2697 (1995).

9. Finally, a large judgment against Cyber Promotions, Inc., did it in. The successful action did not stop spam altogether. Spam companies began to use phony e-mail addresses and even third parties' addresses to protect themselves from "bombing" and legal challenges, which did not always work. For example, in *America Online v. Over the Air Equipment, Inc.*, a Virginia federal judge issued an injunction enjoining the company not to send any more unsolicited e-mail to AOL members. Over the Air Equipment used forged AOL return addresses in its headers. The e-mail supposedly authored by a female contained a

hyperlink to an adult entertainment site, which AOL complained adversely affected its reputation, among other violations. Over the Air Equipment agreed to pay substantial damages. Deborah Branscum, "The Big Spam Debate," *Newsweek*, June 22, 1998, 84, 86.

10. *Central Hudson Gas and Electric Corp. v. Public Service Commission*, 447, U.S. 557, 100 S.Ct. 2343, 65 L. Ed.2d. 341, 6, Med. L. Rptr. 1497 (1980).

11. For filtering solution, see the Electronic Frontier Foundation's archive at www.eff.org (accessed 5 January 2003).

12. See, for example, James Fallows, *Breaking the News: How the Media Undermine American Democracy* (New York: Vintage Books, 1997).

13. "Disney Duels with Time Warner," *NewsHour*, www.pbs.org/newshour/ bb/media/jan-june00/abc_5-2.html (accessed 2 May 2000).

14. *Abrams v. United States*, 250 U.S. 616 (1919).

15. Ben H. Bagdikian, *The Media Monopoly*, 6th ed. (Boston: Beacon, 2000); see also Robert W. McChesney, *Rich Media, Poor Democracy* (Urbana: University of Illinois Press, 1999).

16. Jerome A. Barron, *Freedom of the Press for Whom?* (Bloomington: Indiana University Press, 1973), xiv.

17. *Miami Herald Publishing Co. v. Tornillo*, 418 U.S. 241 (1974). See John D. Zelezny, *Cases in Communications Law*, 2nd ed. (Belmont, CA: Wadsworth Publishing Company, 1997).

18. "French Court Bans Book on Mitterrand's Health," *New York Times*, January 21, 1996. The *Times* reported, "A book about the late ex-president Francois Mitterrand's health by his former doctor was banned this week by a French court, but not until after a first printing of 40,000 copies had already sold out. The court ruled on Thursday in favor of a complaint by Mr. Mitterrand's widow, their two sons, and his daughter. They contended that the revelations in the book, *The Great Secret* by Dr. Claude Gubler, constituted 'a particularly serious intrusion into intimate details of President Francois Mitterrand's private family life, and that of his wife and children.'"

19. Doreen Carvajal, "Book Publishers Worry over Threat of Internet," Howard Besser's Web, March 18, 1996, http://besser.tsoa.nyu.edu/impact/w96/ News/News10/0318book.html (accessed 23 March 2007).

20. Leonard W. Levy, *Emergence of a Free Press* (New York: Oxford University Press, 1985), 9.

21. "Iran Adamant over Rushdie Fatwa," *BBC News*, http://news .bbc.co.uk/2/hi/middle_east/4260599.stm (accessed 12 February 2005). A fatwa is a religious order given by an Islamic clerk regarding a controversial issue.

22. *Miller v. California*, 413 U.S. 15 (1973).

23. "Banned Books Online," http://onlinebooks.library.upenn.edu/bannedbooks.html (accessed 5 May 2003).

24. Arlene W. Saxonhouse, *Free Speech and Democracy in Ancient Athens* (New York: Cambridge University Press, 2006).

25. For "The Index of Forbidden Books," or *Index Librorum Prohibitorum*, see Beacon for Freedom of Expression, www.beaconforfreedom.org/about_database/index_librorum.html (accessed 6 June 2002).

26. See "Recantation of Galileo," University of Missouri, Kansas City, School of Law, www.law.umkc.edu/faculty/projects/ftrials/galileo/recantation .html (accessed 7 June 2001).

27. "History of Censorship," *The New Encyclopaedia Britannica*, Vol. 15, 15th ed. (Chicago: Encyclopaedia Britannica, Inc., 2005), 604. "History of Censorship," Encyclopaedia Britannica Online, 25 March 2007, http://search .eb.com/eb/article-14926 (accessed 25 March 2007).

28. John Stuart Mill, "On Liberty," in *J. S. Mill's "On Liberty" in Focus*, ed. John Gray and G. W. Smith (London: Routledge, 1991), 37.

29. Harold L. Nelson and Dwight L. Teeter Jr., *Law of Mass Communications*, 5th ed. (New York: Foundation Press, 1986), 34. Also see, Frederick S. Siebert, *Freedom of the Press in England, 1476–1776* (Urbana: University of Illinois Press, 1952), 260–63.

30. The Bill of Rights, First Amendment.

31. William Blackstone, *Commentaries on the Laws of England* (Chicago: University of Chicago Press, 1979). Quoted in Nelson and Teeter, *Law of Mass Communications*, 7.

32. *United States v. Cooper*, 25 Fed. Cas. 631, no. 14,865 C.C.D.Pa. (1800).

33. Declaration of Independence.

34. *Schenck v. United States*, 249 U.S. 47, 39 S.Ct.247, 63 L.Ed. 470 (1919).

35. *Schenck v. United States*.

36. Thomas I. Emerson, *The System of Freedom of Expression* (New York: Random House, 1970), 17–18.

37. Zechariah Chafee Jr., *Free Speech in the United States* (Cambridge, MA: Harvard University Press, 1941): "The true boundary line of the First Amendment can be fixed only when Congress and the courts realize that the principle on which speech is classified as lawful or unlawful involves the balancing against each other of two very important social interests, in public safety and in the search for truth. Every reasonable attempt should be made to maintain both interests unimpaired, and the great interest in free speech should be sacrificed only when the interest in public safety is really imperiled, and not, as most believe, when it is barely conceivable that it may be slightly affected. In war time, therefore, speech should be unrestrained by censorship or by punishment, unless it is clearly liable to cause direct and dangerous interference with the conduct of war" (35).

38. 339 U.S. 382, 70 S.Ct.674, 94 L.Ed. 925 (1950).

39. In *Burson v. Freeman*, 112 S.Ct. 1846, 119 L.Ed.2d 5 (1992), the Court upheld a Tennessee law banning political campaigning within a hundred feet of the polling station in order to avoid voter intimidation and election fraud, even though polling places are public forums where speech should be uninhibited. Free and fair election is a compelling interest for the government to protect.

40. *Organization for a Better Austin v. Keefe*, 402 U.S.415, 91 A.Ct. 1575, 29 L. Ed.2d 1 (1971).

41. *Bantam Books v. Sullivan*, 372 U.S. 58, 83 S.Ct. 63, 9 L.Ed.2d 584 (1963).

42. In *Texas v. Johnson*, 491 U.S. 397, 109 S.Ct. 2533, 105 L.Ed.2d 342 (1989), Justice William J. Brennan, writing for a 5–4 majority, said that Texas's interest in preserving the dignity of the flag was not sufficient to override Greg L. Johnson's First Amendment right to burn the U.S. flag as a political protest. He wrote, " If there is a bedrock principle underlying the First Amendment, it is that the government may not prohibit the expression of an idea simply because society finds the idea itself offensive or disagreeable."

43. *Cohn v. California*, 403 U.S. 15, 91 S.Ct. 1780, 1971 U.S. Lexis 32, 29 L.Ed.2d 284.

44. *New York Times*, "US Ruling Makes Libel in Cyberspace Punishable," December 18, 2000, www.uri-geller.com/libel_net.htm (accessed 23 March 2007).

45. Jeffrey R. Elkin, "Cybersmears: The Next Generation," American Bar Association, www.abanet.org/buslaw/blt/bltaug01_elkin.html (accessed 2 September 2001). Also see John Schwartz, "Corporate Case in Ohio Raises Questions on Internet Anonymity," *New York Times,* http://topics.nytimes.com/top/reference/timestopics/subjects/l/libel_and_slander/index.html?query=YAHOO%20INC&field=org&match=exact (accessed 17 October 2000).

46. A. Michael Froomkin, "Anonymity in the Balance," Michael Froomkin's homepage, http://osaka.law.miami.edu/~froomkin/articles/balance.pdf (accessed 23 March 2007).

47. *New York Times Co. v. Sullivan*, 376 U.S. 254, 84 S.Ct. 710, 11 L.Ed.2d 686 (1964).

48. *New York Times Co. v. Sullivan*.

49. *Gertz v. Robert Welch, Inc.*, 418 U.S. 329, 94 S.Ct. 2997, 41 L.Ed.2d 789, 1 Med. L. Rptr. 1633 (1974). In *Chaplinsky v. New Hampshire*, 315 U.S. 568, 62 S.Ct. 766, 86 L.Ed. 103 (1942), the Court held that "fighting words" do not have idea content and therefore do not have First Amendment protection.

50. Nelson and Teeter, *Law of Mass Communications,* 344.

51. *Reno v. ACLU*, available at Epic.org, www2.epic.org/cda/cda_decision.html (accessed 23 March 2007).

52. *FCC v. Pacifica Foundation*, 438 U.S. 726 (1978).

53. *FCC v. Pacifica Foundation.*

54. For more information about PICS, visit the World Wide Web Consortium website at www.w3.org/PICS.

55. 18 U.S.C. 1461 (1999).

56. *Roth v. United States*, 354 U.S. 476 (1957).

57. *Roth v. United States.*

58. *Memoirs v. Massachusetts*, 383 U.S. 413, 86 S.Ct. 1975, 16 L.Ed.2d 1, 1 Med. L. Rptr. 1390 (1966).

59. *Miller v. California*, 413 U.S. 15 (1973). John D. Zelezny, *Communications Law*, 4th ed. (Belmont, CA: Wadsworth/Thomas, 2004), 433.

60. *Miller v. California.*

61. The criminal version of the federal racketeering law is called the Racketeer Influenced and Corrupt Organizations (RICO) Act. The law was originally passed to fight organized crime and drug dealing.

62. *United States v. O'Brien*, 391 U.S. 367 (1968).

63. *Southern Promotions v. Conrad*, 420 U.S. 546, 95 S.Ct. 1239, 43 L.Ed.2d 448, 1 Med. L. Rptr. 1140 (1975).

64. *Campbell v. Acuff-Rose Music*, 510 U.S. 569 (1994), available in the Supreme Court Collection of the Legal Information Institute at Cornell Law School, http://supct.law.cornell.edu/supct/html/92-1292.ZS.html (accessed 7 March 1994).

65. *Fort Wayne Books v. Indiana*, 109 S.Ct. 916, 931n1 (1989).

66. 18 U.S.C. 2251–57.

67. *New York v. Ferber*, 458 U.S. 747, 102 S.Ct. 3348, 73 L.Ed.2d 1113, 8 Med. L. Rptr. 1809 (1982).

68. *New York v. Ferber*. Justices John Paul Stevens and William J. Brennan held in their separate concurring opinions that portrayal of some teenage sexual activity might be artistic, therefore, deserving of First Amendment protection:

Justice Brennan: "In short, it is inconceivable how a depiction of a child that is itself a serious contribution to the world of art or literature or science can be deemed 'material outside the protection of the First Amendment.'"

Justice Stevens: "Moreover, it is at least conceivable that a serious work of art, a documentary on behavioral problems, or a medical or psychiatric teaching device, might include a scene from one of these films and, when viewed as a whole in a proper setting, be entitled to constitutional protection. The question whether a specific act of communication is protected by the First Amendment always requires some consideration of both its content and its context."

69. *Osborne v. Ohio*, 495 U.S. 103, 110 S.Ct. 1691, 109 L.Ed.2d 98 (1990).

70. *FCC v. Pacifica Foundation*, 438 U.S. 726, 98 S.Ct. 3026, 57 L.Ed.2d 1073, 3 Med. L. Rptr. 2553 (1978).

71. *FCC v. Pacifica Foundation*, 438 U.S. 726 (1978).

72. *Ginsberg v. New York*, 390 U.S. 629, 88 S.Ct. 1274, 20 L.Ed.2d 195, 1 Med. L. Rptr. 1424 (1968).

73. *Action for Children's Television v. Federal Communication Commission*, 932 F.2d 1504 (D.C. Cir. 1991).

74. In *Cruz v. Ferre*, 755 F.2d 1415, 1420 (11th Cir. 1985), the Court said, "Cablevision does not 'intrude' into the home. The Cablevision subscriber must affirmatively elect to have cable service come into his home. Additionally, the subscriber must make the additional affirmative decision whether to purchase an 'extra' programming service, such as HBO. . . .These services publish programming guides which identify programs containing 'vulgarity,' 'nudity,' and 'violence.' Additionally parents may obtain a 'lockbox' or 'parental key' device enabling parents to prevent children from gaining access to 'objectionable' channels of programming."

75. *Sable Communications of California, Inc. v. FCC*, 492 U.S. 115 (1989).

76. Marty Rimm, "Marketing Pornography on the Information Superhighway: A Survey of 917,410 Images, Descriptions, Short Stories, and Animations Downloaded 8.5 Million Times by Consumers in over 200 Cities in Forty Countries, Provinces, and Territories," *Georgetown Law Journal* 83 (June 1995): 1849–1934.

Philip Elmer-Dewitt, "On a Screen near You: Cyberporn," *Time Magazine*, July 3, 1995, www.time.com/time/magazine/article/0,9171,1101950703-134361,00 .html (accessed 23 March 2007). Also see Mike Godwin, *Cyber Rights: Defending Free Speech in the Digital Age* (New York: Random House, 1998), 206–59.

77. *Red Lion Broadcasting Co. v. FCC*, 395 U.S. 367 (1969).

78. *Turner Broadcasting System v. FCC*, 129 L.Ed.2d 497, 514 (1994).

79. 47 U.S.C. 223(d). In part the act reads as follows: "Whoever in interstate or foreign communications knowingly uses any interactive computer service to display, in a manner available to a person under 18 years of age, any comment, request, suggestion, proposal, image, or other communication that, in context, depicts or describes, in terms patently offensive as measured by contemporary community standards, sexual or excretory activities or organs, regardless of whether the user of such service placed the call or initiated the communication, shall be fined (up to $250,000) or imprisoned not more than two years, or both."

80. *Reno v. ACLU*, 521 U.S. 844 (1997).

81. *Reno v. ACLU*, 521 U.S. 885 (1997).

82. *ACLU v. Reno* (COPA), 31 F.Supp.2d 473, 495 (E.D.Pa, 1999).

83. *ACLU v. Reno* (COPA), 217 F.3d 162 (2000).

84. See "The First Amendment and the Media," The Media Institute, www.mediainstitute.org/ONLINE/FAM2003/2-c.html (accessed 1 June 2002).

85. *United States v. Thomas*, 74 F.3d 701, cert denied, 519 U.S. 820 (1996).

86. *Ashcroft v. Free Speech Coalition*, No. 00-795, decided on April 16 by the U.S. Supreme Court.

87. Justice Anthony Kennedy, Justice Sandra Day O'Connor, Chief Justice William Rehnquist, "Excerpts from Opinions in Ruling on the Child Pornography Prevention Act," *New York Times*, April 17, 2002, A16, http://query.nytimes.com/gst/fullpage.html?res=9B00E4DA1F3CF934A25757C0A9649C8B63&n=Top%2FReference%2FTimes%20Topics%2FOrganizations%2FS%2FSupreme%20Court%20 (accessed 23 March 2007).

88. "Excerpts from Opinions."

89. "Excerpts from Opinions."

CHAPTER 7

1. For a discussion of how the media and the message interact to create new actualities, see Marshall McLuhan, *Understanding Media* (New York: Signet Books, 1964); Marshall McLuhan and Quentin Fiore, *The Medium Is the Massage* (New York: Simon and Schuster, 1967).

2. Nicholas Negroponte, *Being Digital* (New York: Random House, 1995). Also see Henry Jenkins, *Convergence Culture* (New York: New York University Press, 2006).

3. Jonathan L. Zittrain, "The Generative Internet," *Harvard Law Review* 119 (2006): 1974, www.harvardlawreview.org/issues/119/may06/zittrain.shtml (accessed 3 June 2006). Also see Tom Zeller Jr., "Your Life as an Open Book," *New York Times*, August 12, 2006, www.nytimes.com/2006/08/12/technology/12privacy.html?_r=1&n=Top%2FReference%2FTimes%20Topics%2FPeople%2FZ%2FZeller%2C%20Tom%20Jr.&oref=slogin (accessed 13 August 2006).

4. Harold A. Innis, *The Bias of Communication* (Toronto: University of Toronto Press, 195), 33.

5. "The keyboard is my café," says William J. Mitchell in *City of Bits: Space, Place, and the Infobahn* (Cambridge, MA: MIT Press, 1995), 7.

6. Scott Bukatman, *Terminal Identity* (Durham, NC: Duke University Press, 1993), 105.

7. Reuters, "Japanese Suicide Pacts," People to People Net, http://p2pnet.net/story/3943 (accessed 18 February 2005).

8. "Sheriff: Online Suicide Pact Had Sexual Overtones/Alleged Plot Also Involved People in Canada, Britain," CNN, www.cnn.com/2005/US/02/12/valentine.suicide (accessed 12 February 2005).

9. Barry Schwartz, Hazel Rose Markus, and Alana Conner Snibbe, "Is Freedom Just Another Word for Many Things to Buy?" *New York Times Magazine*,

February 26, 2006, www.nytimes.com/2006/02/26/magazine/26wwln_essay.html?ex=1298610000&en=b3fd339698b5d3b4&ei=5090&partner=rssuser land&emc=rss (accessed 26 February 2006). Also see Barry Scwartz, *The Paradox of Choice: Why More Is Less* (New York: HarperPerennial, 2005).

10. Schwartz, Markus, and Snibbe, "Is Freedom Just Another Word."

11. Schwartz, Markus, and Snibbe, "Is Freedom Just Another Word."

12. Amartya Sen, "Democracy Isn't Western," *Wall Street Journal*, March 24, 2006, A10. Also see Amartya Sen, *Identity and Violence: The Illusion of Destiny* (New York: W. W. Norton, 2006), 51–55.

13. Sen, *Identity and Violence*.

14. "Magna Carta," Britannia History, www.britannia.com/history/docs/magna2.html (accessed 4 June 2005).

15. *Hamdan v. Rumsfeld* (No. 05-184) 415 F.3d 33, reversed and remanded, June 29, 2006.

16. The text of Tolstoy's story "How Much Land Does a Man Need?" is available on the Literature Network at www.online-literature.com/tolstoy/2738 (accessed 23 March 2007).

17. Charles Murray, David Friedman, David Boaz, and R. W. Bradford, "Does Freedom Mean Anarchy?" *Liberty* 18, no. 12 (December 2004), http://libertyunbound.com/archive/2004_12/editors-anarchy.html (accessed 4 May 2006).

18. Joseph Conrad, *Heart of Darkness* (New York: Everyman's Library, 1993).

19. Edwin Black, *IBM and the Holocaust* (New York: Little, Brown, 2001). Also see Peter Reydt, "How IBM Helped the Nazis," World Socialist Website, June 27, 2001, www.wsws.org/articles/2001/jun2001/ibm-j27.shtml (accessed 24 June 2001).

20. John Locke, *The Second Treatise of Government*, ed. Thomas P. Peardon (London: Prentice Hall, 1952).

21. John T. Shawcross, *Complete Poetry and Works of John Milton* (New York: Modern Library, 1950), 719.

Index

120; as a strategic tool, 47;
technology-driven invasion of,
44; and trust in cyberspace,
119–22
private speech: and confidential
conversation, the difference, 62
probable cause, 49, 55, 57, 77, 79,
80, 82–84, 86–87, 89, 192
Proctor & Gamble, 94
profiling: uses of, 97
Progress and Freedom Forum, 121
protecting children: from indecency
and pornography, 193–201
Protection of Children against Sexual
Exploitation Act, 193
pseudonymity, 84
public accommodation, 24
public commons, 107; the golden age
of the, 124
public's right to know, 64
Publius: *Federalist Papers*, 71
pyramids, 30

al Qaeda, 57, 118, 210
Queen (Bloody) Mary I: daughter of
King Henry VIII, 137, 146
Quran, 1, 3

Random House, 144
Ray, William, 64
Reader's Digest Association, 142
Reed, Judge Lowell, Jr., 197
Rehnquist, Chief Justice William, 62
Reidenberg, Joel, 101
Reno, Janet, 109
Reno v. ACLU, 196, 197
Rhapsody in Blue, 140
right to access, 44, 72
right to gather information: under the
First Amendment, 103
right to know, 16, 61, 72, 168
right to opt out, 104
right to reply, 166

Ringley, Jennifer, 7
*Rise of the Virtual State: Wealth and
Power in the Coming Century*
(Rosencrance), 32
Robitaille, The Rev. Dr. Glenn, 68
Roman Catholic Church: the *Index
librorum prohibitorum*, 169
Romeo and Juliet, 127
Roosevelt, Franklin D., 209
Rosencrance, Richard, 32–34
Roth v. United States, 190, 242
Roush, Wade, 221n10
roving wiretap, 83, 88
Royal Air Force (RAF), 59
RSS-feed-generated newspaper, 13
rule of law: testing the, 50–52
Rushdie, Salman, 167
Russia: after the Bolshevik
revolution, 3

*Sable Communications of California
v. FCC*, 195, 243
"safe harbor," 193
Salinger, J. D. 142–45
Salinger v. Random House, Inc.,
235n29
Samuel Morse: and telegraph, 35
Satanic Verses, The (Rushdie), 167
Saturday Night Live, 134
Saudi Arabia, 5, 6, 23, 75, 95, 147,
188, 201, 217
Scalia, Justice Antonin, 52
Schaeffer, Rebecca, 65
Schenck, Charles, 178
Schenck v. United States, 177–78
Schwartz, Paul, 101
Scientific American, 149, 151, 153
search and seizure, 78
search for meaning, 10
Second Amendment, 3
Second Life, 36, 188, 203–4
secularism: of the free marketplace,
150

About the Author

N. D. Batra, the author of *A Self-Renewing Society* and *The Hour of Television*, is a professor of communications at Norwich University, where he teaches media law, ethics, television criticism, and new media and the Internet. He also teaches corporate diplomacy for the graduate program in diplomacy at Norwich University.

He is a columnist for the *Statesman*, a daily newspaper published from Calcutta, India. "Cyber Age," a widely read weekly column that has been published since 1995 in the *Statesman*, is a thoughtful analysis of complex issues and commentary on current affairs. International diplomacy, political affairs, business and corporate culture in the digital age, and communications technology are some of the areas the column covers week after week. Batra has contributed to *La Revue de l'Inde*, Paris, and is a frequent writer for *AsiaMedia*, an online publication of the Asia Institute at the University of California, Los Angeles. He is now working on a new book, *This Is the American Way*. An avid golfer, he also practices yoga and meditation. He blogs about culture, technology, and politics at http://corporatepower.blogspot.com.